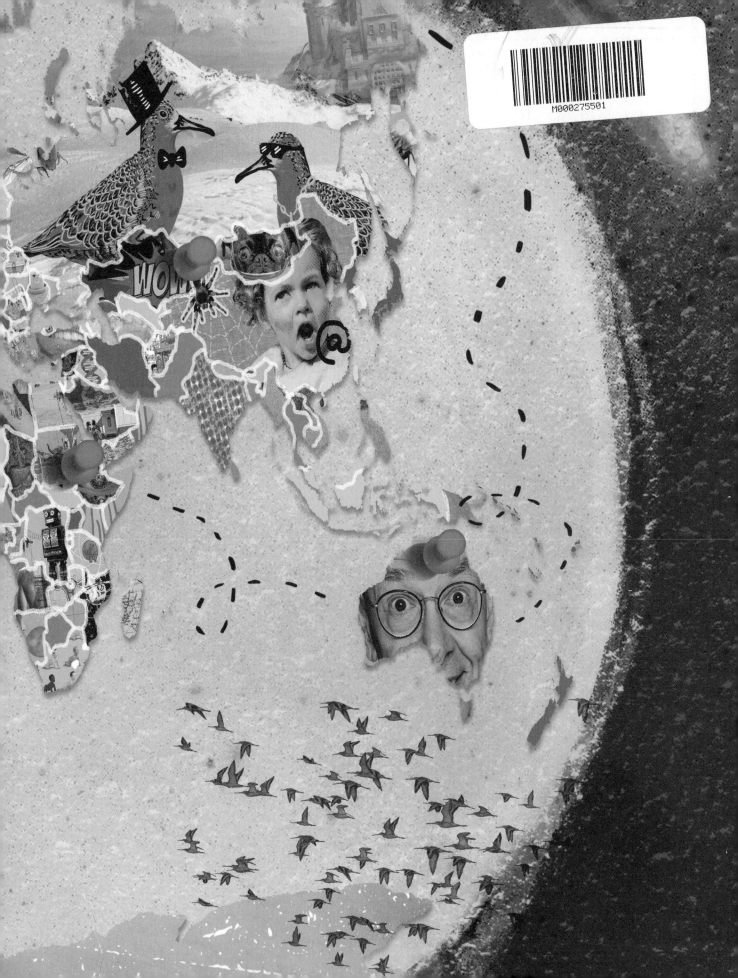

DR KARL'S RANDOM ROAD TRIP

THROUGH SCIENCE

Dr Karl Kruszelnicki

Collages & Illustrations by Pilar Costabal
Design by Lisa Reidy

ABC
BOOKS

CONTENTS

START YOUR ENGINES!   4

Why is Wombat Poo Cube-Shaped?   6

Closest Planet to Earth –
And the Winner is ...   12

Genetic Testing Saves Babies   18

How Do Birds Sleep?   22

Ships & Ocean-Level Rise   28

Hole Phobia?   32

Big Brother Internet?   36

Coral Spawning   44

Tongue Taste Map   50

Balloon Popping   58

Asteroid Dodging   60

Coming Clean on Clothes   64

Why Elephants Don't Get More Cancers   72

Cannibalism – A Nutritious Dining Option?   78

Barcode Invention   82

Alkaline Diet   90

Fat Birds   100

Pregnancy Double Whammy   106

Fire in the Snow – Incinerating Toilet   112

Spaghetti Snapping   114

Vaping & E-Cigarettes –
the Old Bait-and-Switch   120

A2 vs A1 Milk   128

Kiss the Sun   138

Do Fish Drink Water?   148

Human Faults   156

Missing Mass of an Atom   162

Rogue Planets   170

END NOTES   178

PICTURE CREDITS   184

THANK YOU!   186

ABOUT DR KARL   188

GET EXTRA EASTER EGGS WITH THIS BOOK!   190

START YOUR ENGINES!

I'm a rev head from way back, so hitting the road with just my curiosity for company is right up my alley.

And curiosity is all you need on this trip. As Albert Einstein once said, 'I have no special talents. I am only passionately curious.'

Really (?!), I think he vastly underestimated himself. But it's true that curiosity takes you places you could only imagine. It also makes everything in the Universe interesting, which means it's hard to be bored.

Mind you, boredom has a place in the Universe as well. That's because, if you're frantically busy, you have no time for anything – even curiosity. You need time to be creatively bored to let the Big Questions bubble up from deep inside. The trick is to aim for indifferent boredom, where you're relaxed and calm, but not fully engaged with the outside world. A bit like when you go on a lo-o-o-o-ng road trip ...

So big it up for the right type of boredom and a lot of curiosity, and let's make tracks. After all, Einstein also said, 'The important thing is to not stop questioning. Curiosity has its own reason for existing.'*

Have you heard of augmented reality? It's awesome, and in this book are some 'Easter eggs' that can transport you there. For more info, turn to page 190.

*Read more at: brainyquote.com/quotes/albert_einstein_174001

WHY IS WOMBAT POO CUBE-SHAPED?

Life is uncertain — no matter who you are, where you're from, or what you do. But there are some guarantees: you are born and you die, and, in between, you eat and poo. But not all poo is the same.

Look up the Bristol Stool Chart – ideally when you aren't eating – which grades human poo. By the way, 'stool' is a medical word for poo. You'll see that faeces (yet another word for poo) can range in shape from hard little nuggets or pellets, right through to a mushy browny-yellow liquid. The 'variations' in shape and hardness give clues about the health of your gut.

And, yes, this does mean that you should look and check your own poo. In Germany, some of the toilets have a special little elevated platform above the water where the poo lands, so you can inspect it carefully each time.

Given the important information in poo, it's no surprise that some people specialise in studying it. However, one mystery intriguing the specialists – get this – is that wombat poo is shaped like a cube. Yes, it's got six sides, with each side 2 centimetres (cm) long! Why? How does an animal poo out dice!?

WOMBAT 101

Now, before I show you the poo, let me show you the animal that produces it.

The three known wombat species live in southeastern Australia, and in a tiny isolated patch of just 7.5 square kilometres (km^2) in central Queensland. Since the Europeans came, their living space has shrunk by at least 90%. They were branded 'vermin' in 1906, and by 1925, people were being paid a bounty for killing them. We were so ignorant back then. Today, they are officially protected in every state, except Victoria. Since then, attitudes have changed and we have whacked their image onto Australian stamps and coins.

Wombats are chunky little marsupials that dig underground burrows. They have a backward-facing pouch. Because they spend a lot of time digging tunnels, a 'backwards' pouch means less dirt gets into it. They're up to a metre long, and very solid. They weigh between 25 and 40 kilograms (kg). These hefty little animals are mostly nocturnal. They eat predominantly grasses and roots.

And finally – to the shock of many who have studied them – each night they will dump out some 90 little cubical poos, in some 15 bowel movements.

So, let's get to the *bottom* of this wombat poo mystery …

Bristol Stool Chart

Type 1:
Separate hard lumps like nuts (hard to pass)

Type 2:
Sausage-shaped but lumpy

Type 3:
Like a sausage but with cracks on its surface

Type 4:
Like a sausage or snake, smooth and soft

Type 5:
Soft blobs with clear cut edges (passed easily)

Type 6:
Fluffy pieces with ragged edges, a mushy stool

Type 7:
Watery, no solid pieces (entirely liquid)

Wombat boy

Peter J. Nicholson was just a 15-year-old schoolboy in 1960 when he wrote the original study of the bare-nosed wombat.

He was at a boarding school in Victoria, in the foothills of the Great Dividing Range. Late at night, while everybody was sleeping, he'd sneak out a few hundred metres to the wombat burrows, and then slide his skinny body into a burrow. I'd be very scared to do this – wombats are known to kill dogs that invade their burrows by squashing them against the wall of the burrow until they die. But P.J. did it for a whole year, crawling tens of metres into their burrows and studying their secret habits. He was never bitten by a wombat. He could even 'call' to them and they would come to him as if he was a friend.

His focus and dedication were impressive for one so young. Big it up for his school for turning a blind eye to his subterranean explorations. Times sure have changed.

ANIMAL WASTE SCIENCE

Well, thankfully, a group of researchers think they've *cracked* it. (Sorry – another *crappy* pun. 💩)

Dr Patricia Yang, in Mechanical Engineering at the Georgia Institute of Technology in Atlanta, is a specialist in animal wastes. Previously she studied animal urination, and reported that most mammals, regardless of their size, urinate for roughly a 'similar' amount of time – about 21 seconds. (The variation was large – 7 to 34 seconds.) This won her an Ig Nobel Prize in 2015, a very great 'honour'. (These prizes are usually given for research that makes you laugh, and then look and learn. In 2002, I was awarded an Ig Nobel Prize for my groundbreaking research into Belly Button Fluff, and why it is almost always blue.)

WOMBAT POO

Moving along, in 2018, Dr Yang decided to tackle the Big One – cuboidal wombat poo. Every other land mammal has poo that's in round pellets, tubular coils or messy piles. So, what's with wombat poo-cubes?

Wombats are totally vegetarian, and they have a 10-metre (m) gut – as long as ours. Their gut needs to be this long to extract enough foody goodness from the plants they eat. They also have a really slow digestion – some 14 to 18 days. The finished poo at the end is really dry, so it will hold any shape it's given. (By the way, a wombat intestine filled with poo can expand two to three times wider than normal, but a human gut filled with poo expands only a little. Is this a clue to poo production?)

So now let's get back to Dr Yang. A Tasmanian researcher sent Dr Yang the guts from two wombats. The wombats had unfortunately been hit by cars. (By the way, how *do* you send dead animal guts through the mail?) When Dr Yang surgically opened the guts up to study them, the little poo-cubes were there.

She noticed that the cute poo-cubes were located only in the very last few metres of that 10-m-long wombat gut. So, somehow, they were being formed into that cuboidal shape in the bottom of the wombat colon and rectum.

STRANGE SHAPE TO END OF GUT

Now, to wrap your brain around this next bit, just think of a long, skinny balloon.

When you blow up a skinny balloon, it turns into a long cylinder. That's because the balloon rubber has the same elasticity (or firmness) along its whole length. So, when you fill it with air, this creates a regular tube or sausage shape. If you poke your finger into the rubber balloon anywhere along its length, you'll feel the same resistance from the rubber.

From what we know about animal colons, like skinny balloons, we all have one thing in common: there's an even amount of 'firmness' or give on all sides. So, as our poo is gradually manufactured, it gets evenly squeezed from all sides – and what comes out is usually shaped like a tube or a sausage. If you inflated a regular colon, in cross-section it would look like a circle.

But the wombat colon is different.

Suppose you had two lines of weakness running along part of the length of your long skinny balloon, and you blew it up, then it would bulge out along these lines of weakness. In cross-section, our balloon is no longer round like a cylinder. Instead, it would look more like an ellipse.

Sure enough, when Dr Yang looked at the rectum and colon of the wombats, she found something very strange. At the tail-end of the wombat gut, there were two weaker, or less firm, lines running longways – just where the cuboid poo gets shaped. Dr Yang said in *The Guardian*, 'Wombat intestines have periodic stiffness, meaning stiff-soft-stiff-soft, along the circumference to form cubical faeces.'

Now it's a bit of a leap to make from these two longitudinal lines of weakness in the wombat colon and

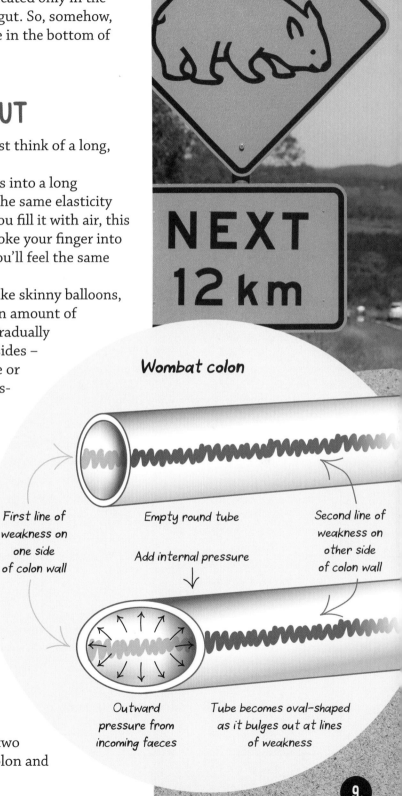

Wombat colon

First line of weakness on one side of colon wall

Empty round tube

Second line of weakness on other side of colon wall

Add internal pressure

Outward pressure from incoming faeces

Tube becomes oval-shaped as it bulges out at lines of weakness

Elephant poo

When I was *squatting* (😊) in Glebe in the inner west of Sydney, I had a small garden for growing vegies. I used compost to nourish the soil. The compost came from my kitchen and that of my neighbours. One day, the circus came to Wentworth Park, just a hundred metres from my house.

I fell in love with elephant poo – for adding to my garden. Each elephant would poo about 50 kg per day, and the circus was more than happy for me to wheel it away in a wheelbarrow. I was surprised and delighted that elephant poo was so light and fluffy. It mushed into my garden really well. That burst of elephant poo improved the soil of my little garden enormously.

rectum all the way to fully cuboidal wombat poo. (If there were four – not just two – longitudinal lines of weakness, it would certainly be a lot easier to understand how wombat poo is shaped into cubes.) But it does give us a hint of how flat surfaces and edges might be formed here. To fully understand the process, we might have to MRI scan a wombat's intestine, while it was in the process of actually producing this wonderful firm cube.

Now, I should also point out that we don't really know *why* wombats have evolved this nifty trick. Biologists tell us that a male wombat tends to keep away from areas where another male wombat has left its poo.

Perhaps the handy cuboidal shape means that the poo won't roll away. This means that a wombat can use its poo a bit like a fence – to mark its territory. But then, why doesn't every animal that marks out its territory have cuboidal poo? Or maybe cuboidal poo is just an accidental side effect of wombats trying to squeeze every drop of moisture out of their food?

While we can make some informed guesses, there are still many unanswered questions around the intriguing subject of wombat poo.

THE SHAPE OF THINGS TO COME

In case you're wondering what's the use of knowing *how* wombats make cuboid poo, here's something to ponder.

At the moment, we humans have only two principal industrial methods to make cubical shapes – we form or mould them into a cube, or we start with some other shape and then just keep cutting stuff away until we end up with a cube.

But wombats seem to use a completely different, third method – one that we still don't fully understand. Just like Albert Einstein's theory of relativity gave us GPS navigation, and Stephen Hawking's black holes gave us Wi-Fi, perhaps wombat poo will give us new engineering processes.

At least wombat poo is not as rare as hen's teeth. With each wombat making 90 little cubes each night, the scientists have plenty to see how the dice tumbles …

WILD GUESSES

Maybe the mysterious pyramids are actually giant alien poo, half buried at an angle in the desert sands?

Maybe wombats have cubical poos because they always eat a square meal?

Maybe wombats are secret gamblers and constantly make their own dice?

And how do the cubes come out – point or flat side first?

And if they come out flat side first, is there a loud bang as the anal sphincter snaps shut?

BAAM!

CLOSEST PLANET TO EARTH –
AND THE WINNER IS...

In our Solar System, we have lots of neighbours, scattered at various distances away. Our nearest neighbour is, of course, the Moon, but it's not a planet.

Around the middle of our Solar System, there's the Big Guy – the Sun, also not a planet. But now we get to the eight planets – Mercury being closest to the Sun and Neptune being the furthest from it.

(Sorry, Pluto, I still miss you.)

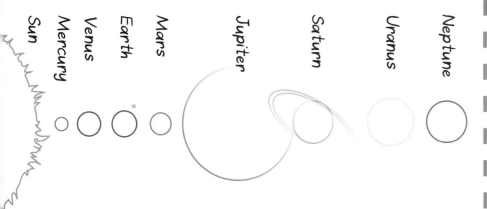

Sun Mercury Venus Earth Mars Jupiter Saturn Uranus Neptune

And here's a very specific and mathematical question: averaged out over half a century, which planet spends most of that time being our nearest neighbour?

Since our home, Earth, is the third rock from the Sun, I'd always thought that our nearest neighbour would be either the second rock from the Sun (Venus), or the fourth rock from the Sun (Mars). (I leant towards Mars.)

But I was 100% wrong. How's that for surprising, or as the scientists like to say, 'counter-intuitive'?

PLANETS RESONATE WITH ONE ANOTHER

Gravity is the weakest of the four known Fundamental Forces, but it does reach to the end of the Universe. And it never switches off – it's always quietly doing its sucky-bending-space thing.

In our Solar System, the Sun is obviously the Biggest Guy, with the biggest mass and the biggest gravity. Venus and Earth, both small players, have a special gravitational and mathematical relationship with each other. It helps being next-door neighbours. They have an 8:13 relationship with their orbits around the Sun. In other words, in the time the Earth does 8 orbits, Venus does 13. Their orbits are not independent of each other – they are linked by this 8:13 relationship.

And way out away from the Sun, Pluto and Neptune have their own version of linked orbits. Theirs is a 2:3 relationship. In the time that Pluto has done 2 orbits, Neptune has done 3. And they're always tugging at each other to keep this 2:3 relationship exact. (Even though Pluto has been demoted from planet status, it's still an influencer. It should probably have its own social media accounts. 😊)

DR KARL'S Q+A

Why are we so special?

We know of over 1,000 exoplanetary systems (other solar systems) in the Milky Way. They're very different from our Solar System. Blame Saturn.

Our Solar System has four rocky planets close in (Mercury, Venus, Earth and Mars) and four big gas planets (Jupiter, Saturn, Uranus and Neptune) further out.

But most of the exoplanetary systems show two different patterns. The first pattern has so-called super-Earths – heavier than Earth, but lighter than Neptune. We've got no planets like that. Super-Earths are usually very close to their star.

The physics/maths tells us that when a star system evolves, you end up with a bunch of super-Earths really close in. Or sometimes you start with a big Jupiter and a bunch of super-Earths close in. The physics/maths further tells us that the Jupiter planet migrates inwards and acts like a wrecking ball, destroying the super-Earths and leaving Jupiter on its own.

The second pattern has an enormous Jupiter-sized planet – but usually, so close to its star, that it's called a hot Jupiter.

But in our Solar System, we (unusually) have Saturn. It migrated inwards with Jupiter, then dragged it out again. The debris left over gave us the four small rocky planets – including Earth – making our Solar System very special.

INNER SOLAR SYSTEM 101

So, let's go back to basics.

We'll ignore the outer planets – the four big gas planets, Jupiter, Saturn, Uranus and Neptune.

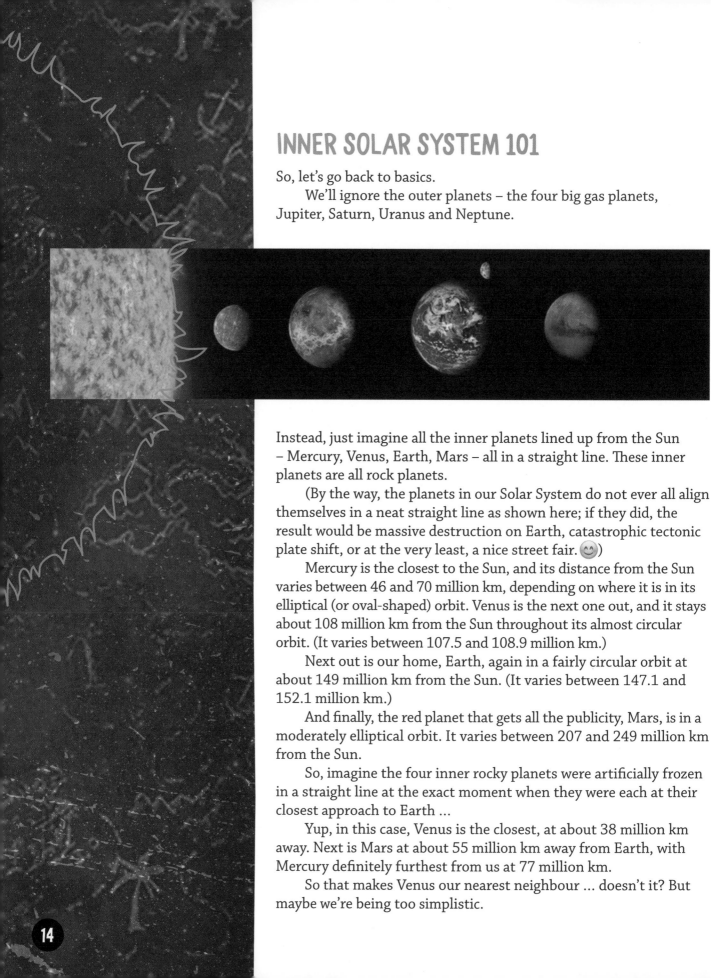

Instead, just imagine all the inner planets lined up from the Sun – Mercury, Venus, Earth, Mars – all in a straight line. These inner planets are all rock planets.

(By the way, the planets in our Solar System do not ever all align themselves in a neat straight line as shown here; if they did, the result would be massive destruction on Earth, catastrophic tectonic plate shift, or at the very least, a nice street fair. ☺)

Mercury is the closest to the Sun, and its distance from the Sun varies between 46 and 70 million km, depending on where it is in its elliptical (or oval-shaped) orbit. Venus is the next one out, and it stays about 108 million km from the Sun throughout its almost circular orbit. (It varies between 107.5 and 108.9 million km.)

Next out is our home, Earth, again in a fairly circular orbit at about 149 million km from the Sun. (It varies between 147.1 and 152.1 million km.)

And finally, the red planet that gets all the publicity, Mars, is in a moderately elliptical orbit. It varies between 207 and 249 million km from the Sun.

So, imagine the four inner rocky planets were artificially frozen in a straight line at the exact moment when they were each at their closest approach to Earth …

Yup, in this case, Venus is the closest, at about 38 million km away. Next is Mars at about 55 million km away from Earth, with Mercury definitely furthest from us at 77 million km.

So that makes Venus our nearest neighbour … doesn't it? But maybe we're being too simplistic.

COMPLEXITY IS COMPLEX

A system is much more than the sum of its parts, to paraphrase an old saying. For example, a rainforest is much more than just a mix of dirt, insects, animals and trees.

Another example is a car engine. I have built a car engine from its individual parts – crankshaft, conrods, pistons, valves, fuel pump, distributor, etc. When you've finished assembling it, it just sits there – dead. But give it fuel and electricity, and start it spinning, then something almost miraculous happens. The engine comes alive! Give it more fuel, and it will spin faster.

The Solar System is also more than just a bunch of planets – and much more than the sum of its parts.

For example, one of the planets has somehow evolved an intelligent life form on its surface! And now this life form is looking for other life forms elsewhere in the Solar System.

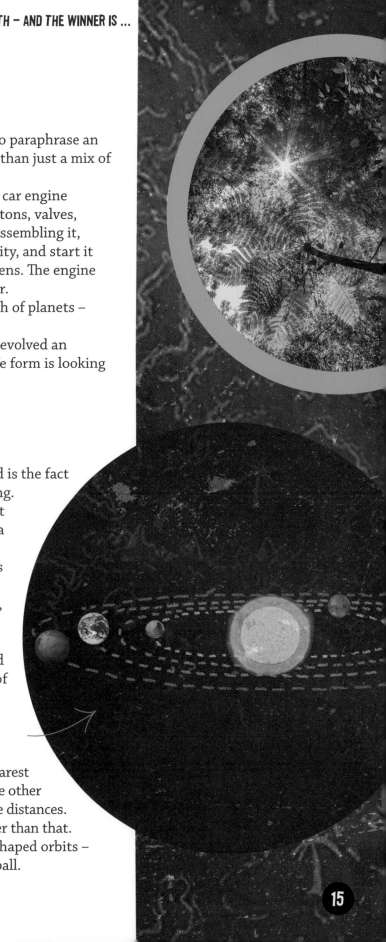

DYNAMIC SOLAR SYSTEM

Well, one major consideration that we have ignored is the fact that in the real Solar System, the planets are moving.

For example, the inner planets orbit the Sun at different rates. Mercury takes 88 Earth days to do a complete orbit of the Sun, Venus takes 225 days, Mother Earth takes 365-and-a-bit days, while Mars takes 687 days.

Because the planets move at different speeds, Venus could actually be the inner planet furthest away from Earth. This happens every now and then when Venus is on the far side of the Sun, and Mars and Mercury are close to Earth on our side of the Sun.

The celestial objects illustrated here are (from left to right): Mars, Earth, Mercury, the Sun and Venus.

So, to work out which is (on average) Earth's nearest neighbour, you have to carefully work out each of the other planets' exact location, on each day, and measure the distances. But, actually, working out the location is even trickier than that.

You have to take into account their different-shaped orbits – from nearly circular to quite elliptical like a rugby ball.

You also have to take into account how fast they move. When the planets are close to the Sun, they move rapidly; on the other hand, they move more slowly when they are far from the Sun. So if you consider just one of the rocky planets, you'll see that it covers a different distance in each 24-hour period.

Another factor is that the four inner rocks are not all orbiting in the same plane. Relative to the Earth's orbit, Mercury's orbit is quite tilted at 7°; Venus's is about half that at 3.4°; while Mars's is the least tilted at 1.9° relative to the Earth. So, the other rocky planets are sometimes above the orbit of the Earth and sometimes below it. And this, too, slightly affects their distances from us.

Luckily, we have computers today to do all the hard number-crunching. Back in Ye Olde Days, it was just pen on paper, used by people referred to as 'computers' – most of whom were women!

DO THE NUMBERS

But enough of guessing and inferring – let's do the numbers.

Data scientist, Oliver Hawkins, wrote some computer code to help find the answer. It used the actual locations of the inner four rocky planets as they orbited around the Sun, on each and every day for the last 50 years. This is the best way to find out which planet is closest to us, most of the time.

So, which is it?

(Time to open the envelope, and announce the Winner.)

It's ... Mercury!

Yep, Mercury, the first rock from the Sun. Mercury has been our closest neighbour 46% of the time over the last 50 years. Venus was second at 36% and Mars third at 18%.

That is definitely counter-intuitive. I was astonished – but the numbers do not lie.

How come?

Well, Mercury never really gets very close to us. But it never gets very far from us either, because its orbit is the closest to the Sun. So, if all three planets were on the other side of the Sun to us, Mercury would be closer to us than Venus or Mars.

Everybody needs good neighbours. Our closest neighbour is Mercury, but sadly, we probably won't find any good friends there.

An illustration of Mercury's interior based on 2019 research that shows the planet has a solid iron inner core.

Mantle

Crust

Solid inner core

Outer molten core

SHOULDN'T IT BE MARS?

Mars is often casually described as *'our closest neighbour'*. We've sent lots of space rockets to Mars, and we've made plenty of movies about it, which means that Mars gets more than its share of prime-time news.

And, beyond Earth, Mars is probably the planet on which humans would find it easiest to survive.

It has a slightly more comfortable temperature range (–143°C up to +35oC) than the other planets offer, and it has a bit of atmosphere (about 170 times less than ours, but at least it's not a total vacuum).

I strongly suspect that Mars might even have some kind of life on it (but so far we have zero proof).

You can forget about living on Venus in the short term. Thanks to lots of carbon dioxide and a runaway Greenhouse Effect, Venus has a surface temperature around 462°C, and an air pressure around 92 times greater than ours on Earth.

Mercury, even though it orbits closest to the Sun, is not the hottest planet (that title belongs to Venus). Unfortunately, it has virtually no atmosphere at all that could sustain human life.

GENETIC TESTING SAVES BABIES

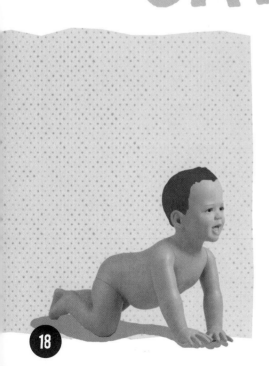

Most of us have watched *CSI*, or have heard of DNA. So, we have a vague idea that DNA can be used in forensic testing to find bad dudes, or for finding long-lost relatives.

Perhaps you've even heard of people working out their risk of inheriting cancer and other diseases, from their DNA. But it's only very recently that our knowledge has advanced to the stage where DNA can be used in a hospital-emergency setting.

DNA testing has now (in a limited way) been used to save the lives of newborn infants suffering from a variety of genetic diseases.

GENETIC DISEASES IN NEWBORNS

On average, genetic disorders and congenital abnormalities appear in about 6% of new-born babies. Usually, these changes do not affect their immediate or later health.

However, it's a very different picture entirely in the Neonatal Intensive Care Unit (NICU). NICU looks after really sick babies. Some 14% of the babies in NICU have a serious genetic disorder. These babies stay in hospital longer than babies without genetic disorders, and need lots more resources.

Sadly, genetic diseases are a major cause of deaths in an NICU.

DNA 101

First, some background. DNA is a very long molecule found inside all of the cells of our body (except for red blood cells). Imagine a ladder twisted around itself into a spiral shape.

If you uncurled and fully stretched out this DNA 'Ladder of Life', it'd be about 2 or 3 m long, with three billion rungs squished in as the tiny 'steps' on the ladder.

One of the great discoveries of the 20th century was how DNA made amino acids. It took half a century to work out the full process. First was the realisation that each group of three rungs formed a code. We then unravelled how this code could command the biological machinery inside a cell to make an amino acid. Bingo 👍. Put enough amino acids together, and you get a protein – and if you put enough proteins together, you get a human being.

So in one sense, DNA is a blueprint for building a living creature – be it a bacterium, a fungus, a dinosaur or a plant.

Adenine

Thymine

Cytosine

Guanine

Sugar-phosphate backbone

DNA

DNA TODAY

This DNA map was soon being used everywhere. Forensic testing, molecular medicine, virology, biofuels, agriculture and animal husbandry, anthropology, evolution and even some cosmetic products became DNA driven.

But to do a full and accurate DNA mapping of one individual person was still expensive and slow.

General genetic 'variations'

I personally have at least three genetic mutations (or 'variations', to be more polite). Medical people are good at finding their own 'variations'.

First, on my upper jaw, in the middle, instead of having four incisor teeth, I have only two. I do have the two central upper incisors, but not the two lateral incisors. This gives me a 'cute' gap between my upper incisors. I inherited this from my mother.

Second, from my father, I have little lipomas scattered across my body in my skin. Lipomas are little flattened 'bumps' of fat, 3–10 mm in diameter.

Finally, in my right shoulder, right at the tip, I have 'incomplete fusion of the right acromial tip'. In plain English, this means that not all the cartilage at the tip of my shoulder turned into bone.

None of these genetic abnormalities are dangerous, and they have not really bothered me. Almost certainly, I have a bunch more that I haven't found yet.

Finally, in 2018, DNA sequencing became cheap enough and fast enough that it could now be used in medical emergencies.

The cost of mapping a person's entire DNA has dropped from $3 billion to just under $10,000. The time to get the results has been reduced from a decade and a half to a few days – and if you push really hard, to less than a day.

So here's the very uplifting news.

We can now use DNA mapping technology to diagnose, and sometimes treat, desperately ill infants suffering from rare, genetically inherited conditions.

We currently know of more than 8,000 genetic conditions that can make you very unwell – such as Angelman Syndrome, Shone's Complex or Wolff-Parkinson-White Syndrome. Sometimes the diagnosis is fairly straightforward – especially if the child is otherwise well.

But sometimes it can be very tricky.

SICK BABIES – PRE DNA TESTING

A diagnosis can be especially urgent, but difficult to obtain, when a newborn baby with a very rare genetic condition is desperately ill in an intensive care unit.

Much of Ye Olde School way of diagnosing many genetic illnesses depended on recognising a pattern in a person's appearance. This might include facial features (e.g. Down Syndrome), or a high palate in the mouth combined with very long spindly fingers (Marfan Syndrome). This body of knowledge was built up over a period of centuries.

Chromosome testing helped sometimes. For example, if there were three chromosomes at Position 21 instead of the usual two, this confirmed the diagnosis of Down Syndrome. But chromosome testing was a 'big picture' diagnostic tool, and couldn't give the fine detail of modern DNA testing.

Getting a diagnosis from the babies' appearance was even harder when they were sick in NICU and fighting for their life. A tiny infant might be hard to get a good look at under all the pipes, tubes and insulating blankets. Their face would show stress and would not be 'normal'.

All of this made diagnosis very difficult, so the advent of cheaper and faster genetic testing was a game-changer.

SICK BABIES – POST DNA TESTING

Recently, two teams in the USA and the UK were able to quickly map the entire DNA of some 60 very sick kids in intensive care with unknown illnesses. For some of the kids, the DNA information was complete enough, and obtained soon enough, to be useful for their management.

Getting the entire genome of the infant was the first step. But then it took at least another four days for the staff to tease out a diagnosis from the massive amounts of data.

The teams got useful diagnoses in about half of the children. These diagnoses included a hormone disorder, a rare mutation that always leads to recurrent kidney cancers, and vascular Ehlers-Danlos Syndrome, which weakens internal tissues (in this infant, causing a ruptured spleen).

There were many reasons the doctors could not get a diagnosis in the other half of the children. We don't have a full genetic snapshot of many genetically transmitted diseases. Sometimes, the genetic differences are both subtle and spread across the entire DNA (not conveniently concentrated in a single location). And the diagnosis can be trickier again because a genetic disease can be very severe in one person yet mild in another – and we still don't know why.

But having a diagnosis is different from having a cure. Overall, about 10% of the kids were able to get treatments that improved their health. This figure is not as good as 100%, but it's a lot better than 0%.

Genetic testing is still expensive. But Neonatal Intensive Care costs even more dollars – and what you're aiming for is a healthy baby with a whole lifetime ahead of them.

DNA testing does save money overall. With genetic testing, they were able to get the kids out of hospital sooner. In the USA, hospital costs were reduced by US$128,000.

With these DNA testing advances, we're getting closer to the day when genetic testing can be done cheaply and on-the-spot, when you visit your GP.

And from there? Who knows. But it is clear we'll keep climbing the rungs of knowledge on the Ladder of Life. After all, scientific curiosity is in our DNA.

And maybe this will be the one test to rule them all …

The Human Genome Project

The Human Genome Project was huge. (A 'genome' is the genetic material of an organism.) This was an international scientific project set up to map every single rung on our biological Ladder of Life – our DNA. This $3-billion project was formally started in 1990 by the United States Department of Energy and the National Institutes of Health – with assistance from geneticists in the United Kingdom, France, Australia, China and elsewhere.

But where did they get the DNA samples that were used in this initial work? They came from a handful of people. So the final 'human DNA map' is actually a mosaic formed from the DNA information gathered from several people – not just one single person.

In the year 2000, the US President Bill Clinton and the British Prime Minister Tony Blair announced that the very first rough draft of most of the human DNA had been completed. But it took until 2006 before the full DNA sequence was published. This sequence is now freely available on the Internet to all researchers.

zzzz

HOW DO BIRDS SLEEP?

Birds do plenty of chirping and warbling that we notice in the daytime. We hear and see them while we and they are both awake, but we hardly ever see them when they're asleep.

So here's the question – how do sleeping birds lie?

Well, they usually hide – but often, they let only half their brain go into a deep sleep. The other half is resting, ready for instant action.

NEST MYTH?

Most of us have the mental image that, after a hard day of chasing worms and warbling beautiful songs, our lovely little bird will retire to its nest. There, it will pull up a cosy little blanket (maybe a flexible leaf), fluff up its feathers (to generate some air insulation for the night), and then (very cutely) burrow its head into the feathers around its neck – and sleep peacefully until the next morning.

Nope, the 'nest' bit is mostly a myth. (Sorry to destroy a lifetime of cartoon imagery!)

Surprisingly, nests are for the *chicks* to hatch and grow up in. Birds hardly ever sleep in nests unless they're babies, or it's a cold night and the adult parents need to cuddle up to their babies to keep them warm. (However, there are a few tropical birds that do have 'dormitory' nests to sleep in.)

By the time the chick is grown, the nests are truly trashed. They are full of bird poo, spilt food, random parasites and sometimes, sadly, a dead chick – so the owners simply abandon them.

At least our mental image is right about one thing. Most birds do tuck their bills into their plumage while they sleep. This gives them access to air that is heated by their bodies, so they breathe warm air (instead of the cold outside air).

What is sleep?

That's tricky.

One pretty good (but not perfect) definition claims that sleeping animals will:

1) have a quieter, or more calm, behaviour than otherwise (e.g. humans move more when they're awake than when asleep);

2) take on a fairly specific posture that most other members of that species will also use (e.g. humans usually lie down flat);

3) need more stimulation than normal to get a response (e.g. you need to talk loudly to get a sleeping person to wake up);

4) wake up if you stimulate them enough (e.g. a sleeping person will eventually wake up if you shout loudly enough);

5) need more sleep after they have been deprived of sleep. Errr ... yes ... unfortunately true ...

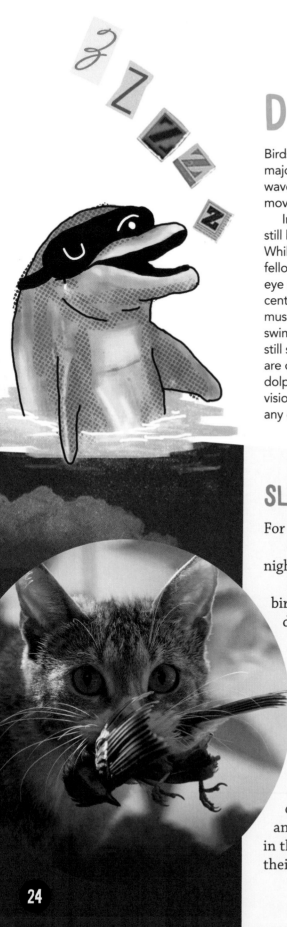

DOLPHIN SLEEP

Birds and mammals have two major types of sleep – SWS (slow wave sleep) and REM (rapid eye movement sleep).

In SWS, a bird or mammal can still have some muscle tone left. While in SWS sleep, dolphins (a fellow mammal) will sleep with one eye open – usually aimed at the centre of the group. Because their muscle tone allows them to keep swimming while asleep, they can still see what the other dolphins are doing, and so keep up with the dolphin pod as it swims along. Their vision will also enable them to locate any external threats.

In REM sleep, however, you generally lose muscle tone. The advantage is that you won't act out your dreams, which usually occur during REM sleep. So you won't run around the bedroom, fighting a dragon. The disadvantage is that you may find it difficult to maintain your temperature, as you stop panting or shivering.

Do sleeping birds and mammals that keep moving while they sleep still have REM sleep, then? We don't know, as there's still lots of missing information about how some birds and mammals can move while asleep.

SLEEP – SAFETY AND WARMTH

For birds, sleep is all about safety and warmth.

Most birds sleep at night. But a few nocturnal birds (owls and nighthawks, for example) are, by definition, most active at night.

The technical term for 'a bird going to sleep' is 'roosting'. How birds roost depends on what kind of bird they are, what they're doing (migrating or simply hanging around), the season, the climate, the weather and so on.

Of course, the top bird predators – such as owls, hawks and eagles – are pretty safe so long as they stay off the ground, away from dogs and cats. (Unfortunately, in Australia, cats kill about a million birds each day.)

But the majority of birds are not powerful predators.

More than half of bird species are perching birds from the Order Passeriformes – sparrows, jays, warblers, cardinals and so on. They mostly sleep in dense vegetation. Around dusk, they find a branch, hook onto it with their clawed feet, and squat down. These 'passerines' have special flexor tendons in their legs that automatically tighten onto the branch. So long as their legs are bent, they are physically locked onto the branch –

The technical term for 'a bird going to sleep' is 'roosting'.

and this action takes no muscular work at all. This grip can be so tight that some birds have been seen sleeping upside down. Once the bird wakes up and straightens its legs, the flexor tendons loosen, and the bird can fly away.

If they are a smallish songbird, they won't sleep on the ground, because a cat will get them. And they won't sleep on an exposed branch, because an owl will get them. So they hide, beautifully camouflaged inside dense brush or behind foliage.

Waterfowl (such as ducks and geese) can't sleep in trees, because they have webbed feet. They are too slow and clumsy at taking off to safely sleep on the ground. So they might sleep on a small island, or on the water. At night, out in the open, they're safe from hawks and eagles, because these predators also sleep at night. The waterfowl can sense vibrations in the water emitted by predators swimming towards them.

Bigger waterfowl (such as flamingos and herons) might roost in the shallows, and again rely on vibrations in the water to warn them of an attack.

Birds such as quail and grouse are unfortunately not very good at flying – and are also fat and yummy. So virtually every creature in the animal kingdom wants to eat them. These delicious birds hide in the densest camouflage they can find.

But what if there's no vegetation, only snow? Well, birds such as the white-tailed ptarmigan rely on their white plumage to merge invisibly into the snow.

Other birds, such as crows, swallows, swifts and starlings, all share amazing communal roosting behaviours within their own groups (even though they are not closely related). Some of them gather in enormous flocks – usually at dusk – to seek safety in numbers.

25

BIRDS SLEEP WITH HALF A BRAIN

Many birds have another trick – they can switch off half their brain while they sleep.

We humans talk about having just one brain inside our skull, but we actually have two separate brains (called hemispheres). We have a brain on the right side (with 43 billion neurons or nerve cells) and another on the left side (again with 43 billion neurons). These separate brains are joined only by a tiny bundle of nerves called the corpus callosum. This is about 10 cm long, and carries only a quarter of a billion nerve fibres. Each of our eyes sends information to both the right and the left brain.

Birds also have two brains, but the eye–brain set-up is different. The information from one eye goes only to the brain on the other side. So the left eye sends information to the right brain, and vice versa.

To get a safe night's sleep, birds close one eye and switch off one brain – and leave the other eye (with the corresponding brain) fully awake and alert.

In fact, they can even turn half-brain sleeping on and off, depending on the circumstances. So, in a large flock of geese roosting on a lake, the geese in the centre of the flock might have both brains asleep. But the more vulnerable birds (on the perimeter) might have one eye open, with its corresponding brain alert to look for predators.

Our little feathered friends can sleep peacefully through the night – resting their bird brains in readiness to again spread their wings.

10 cm

Corpus callosum

MIGRATING BIRDS

We haven't been able to determine if migrating birds can switch off half of their brain, to sleep.

But think about the bar-tailed godwit. It spends over a week flying more than 11,000 km from western Alaska across the Equator, without landing. How does it manage this without sleeping or resting somehow while in flight? Perhaps switching off half of its brain is part of the answer.

Humans sleep with one brain?

We humans will sometimes do the 'sleeping-differently-on-each-side-of-the-brain' thingy.

This can happen when we sleep in an unfamiliar environment, such as in a hotel in a city we've never visited before.

A 2016 study (done in a laboratory, not a hotel) found that, on the first night, the right brain seemed to sleep deeply. But the left brain showed EEG activity indicating a shallow sleep, increased alertness and easier waking from sleep. This so-called 'first-night effect' might act to keep you just a touch more vigilant in a potentially hostile environment. The effect faded away on the second night.

But the study had only a small sample size, and did not specifically separate out left-handers from right-handers.

However, if you travel a lot for work and if this first-night effect is significant for you, then it's probably worthwhile to book rooms in the same hotel chain. The rooms are sometimes reasonably similar – and you get Frequent Stayer points you can cash in! 👍

You need some upside for not getting to sleep in your own bed.

27

SHIPS & OCEAN-LEVEL RISE

If you plonk yourself into a bath full of water, the water level will rise – and overflow. Pretty obvious.

So, what happens to the oceans when we plonk enormous numbers of big ships into them? The ocean level rises – but only by the thickness of a spider web. And this is less than the daily increase in sea level from Climate Change.

OCEAN-LEVEL CHANGE 101

First, let's start with some deep history. The levels of the oceans have varied enormously since our planet formed some 4.5 billion years ago.

Billions of years ago, the first oceans formed from water from different sources – from the atmosphere, from the interior of the Earth, and from that brought in by comets.

Moving forward into the last billion years, our planet seems to have had sets of Ice Ages every 150 million years. These sets of Ice Ages lasted for several million years. Each set of Ice Ages was made up of a series of repeating shorter, or little, Ice Ages.

When I say 'little Ice Age', it was still pretty major. The ice was a kilometre thick over New York, and glaciers covered much of the land surface. The water to make this ice had to come from somewhere – the oceans. That's why, during an Ice Age, the sea level actually dropped by around 100–125 m – compared with today's levels. At the other extreme, when the Earth heated up and there was no ice anywhere on the surface of the land, the sea level rose by about 70 m – again, compared with today's levels. This massive ocean-level rise last happened about 50 million years ago.

The most recent 'pulse' of little Ice Ages began about 3 million years ago. Over the last million years, a little Ice Age would last for about 100,000 years (called a 'glacial period'), then there would be a gap of about 20,000 years when most of the ice melted (called an 'interglacial period'), and then there would be another little Ice Age (another glacial period) for another 100,000 years – and so on for the last million years.

Getting closer to today, the last little Ice Age finished about 12–14,000 years ago. The Indigenous Australians have accurate stories going back some 10,000 years, describing the changes as the ocean levels rose.

DETOUR

Weird coincidence

The mass of the entire world's shipping fleet is a bit over 2 billion tonnes.

The biomass of all the fish in the oceans is about 2 billion tonnes.

This marine fish biomass has dropped by about 80% over the last century.

So if we allow for a bit of error in these figures, and that shipping mass will probably increase while fish mass will probably drop, we arrive at a sad data point.

Sometime quite recently, or sometime really soon, there was/will be more mass of ships in the oceans than of fish.

Sometime quite recently, or sometime really soon, there was/will be more mass of ships in the oceans than of fish.

30

MEASURING OCEAN LEVEL

We have been directly measuring sea level for a few centuries.

In the early days, this was done with tide gauges. These are basically large pipes sitting in the water, with a float inside that bobs up and down as the water level changes. Of course, you have to account for local factors, such as that in some parts of the world, like Alaska, the land is rising, while elsewhere it is falling. There are other factors that affect the local sea level, such as differences in water temperature, wind, ocean currents and atmospheric pressure. More recently, to get global measurements, we have used a series of orbiting satellites, beginning with *TOPEX/ Poseidon*, launched in 1992.

During the whole of the 19th century, the sea level rose by around 6 cm worldwide.

But in the much shorter period of 39 years between 1979 and 2018, the ocean levels rose by about 17 cm – thanks to the ice melting just in Antarctica.

Worldwide, the rate of ocean-level rise is increasing, due to extra greenhouse gases trapping heat. Each day, this heat is equal to the total energy output of exploding 400,000 Hiroshima bombs. That's a lot of extra heat as compared with pre-industrial heat trapping.

(Yes, Global Warming is real, and we humans caused it. There have been previous episodes thousands and millions of years ago, but we humans have caused, and are still causing, this current episode.)

One major cause of this rise in ocean level is that solid ice that was previously on land has been melting, turning into liquid water, and then flowing into the oceans. A 2019 study looked at Antarctic melting over different time periods. In the period from 1979 to 1980, the average mass of ice melting was 40 billion tonnes per year. But from 2009 to 2018, the average had jumped to 252 billion tonnes per year. That's more than a six times increase – and in a very short time. Even worse, that rate of melting is accelerating.

The other major cause of the sea-level rise is thermal expansion. Warm water, after all, takes up more volume than cold water – and yes, the oceans are warming.

But besides the gradual increase in ocean level, there are very strong possibilities of sudden jumps. (The technical term is Positive Feedback Loop – look it up on Wikipedia.) If you're interested, read up on the Thwaites Glacier in West Antarctica. (It's a bit scary.)

So that's the important background stuff out of the way. But what about the effect of shipping, on the ocean level?

SHIPPING

To know what ships do to sea level, you need to measure the ships – and it turns out that there are many different ways to do this. Surprisingly, the two words 'gross tonnage' have nothing to do with the weight of a ship. Instead, they relate to its internal volume! (Who knew?)

What we really want is the total weight of the ship, fuel, crew and cargo. Luckily, Randall Munroe, the author of *XKCD*, that wonderful webcomic of 'romance, sarcasm, math, and language', has done all the numbers for us.

His sums show that the total weight of all the ships floating in the oceans works out to a bit over 2 billion tonnes. By the way, over two-thirds of this weight are oil tankers and cargo ships carrying huge amounts of dirt (coal and iron ore). Navy ships and recreational vessels? They don't add up to much.

Some 2,000 years ago, Archimedes showed that a floating object displaces a weight of water equal to its own weight. That means that floating ships weighing two billion tonnes push aside about two billion tonnes of water. When you evenly spread this amount of water over the surface area of all the oceans on Earth (about 360 million km^2) you end up with the sea level rising just 6-millionths of a metre (or 6 microns).

Now, that sea-level rise is truly microscopic – about the thickness of a spiderweb (or 6 microns).

So yes, being precise and pedantic, if you removed all the world's ships and put them on dry land, the level of the oceans would drop – but only by a tiny amount.

On the other hand, the current level of ocean-level rise due to Climate Change is about 3.3 mm per year. That's about nine microns each day.

So, in less than a day, about 16 hours, Climate Change would bring the water back to its original sea level.

And the worry about ships causing sea-level rise – well, that's just a storm in a teacup.

Bumpy sea level

Surprisingly, melting ice sheets can sometimes lower (not raise) the water around them!

That's because the oceans are not like a small bathtub, with the water level the same everywhere. For example, winds and currents can push up water in certain areas into a gentle swelling.

The enormous amounts of ice on land in Antarctica and Greenland actually increase the local sea level (microscopically) by the action of their gravitational mass pulling the water towards them. As this ice on land melts, the gravitational pull is slightly less, so the local water can flow away. In this scenario, the ice can melt and the local ocean level can drop.

HOLE PHOBIA?

You've heard of phobias, right? They're those strong, irrational fears of something that is probably harmless.

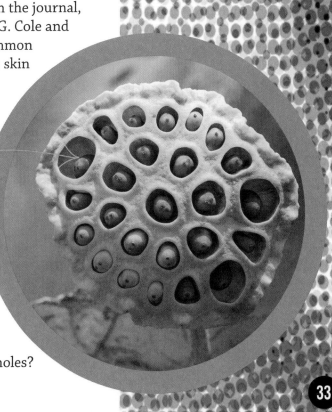

You'll have heard of arachnophobia (which is a fear of spiders); and maybe even triskaidekaphobia (fear of the number 13).

But, what about trypophobia? This is the name for the fear of irregular patterns or clusters of small holes (or shapes).

And here comes the Big Twist: trypophobia might not even be a phobia at all!

TRYPOPHOBIA 101

The word 'trypophobia' was coined around 2005 by someone in an online forum, for people who all experienced a dread of repeated little holes. 'Phobia' is 'a fear of something', while 'trypa' is a Greek root meaning 'holes'.

The symptoms can range from very mild to very severe. Typically, during an episode, a person with this obscure phobia will suffer feelings of fear and helplessness. They might feel their stomach churning or their heart racing. They might shudder, sweat and feel their skin crawl. It might be bad enough to turn into a panic attack. One sufferer said, '[I] can't really face small irregularly or asymmetrically placed holes, they make me like, throw up in my mouth, cry a little bit, and shake all over, deeply.'

Trypophobia got its first scientific nod in 2013 in the journal, *Psychological Science*. The authors of the paper, Geoff G. Cole and Arnold J. Wilkins, simply wanted to find out how common trypophobia was. One of the authors, Cole, had some skin in the game (so to speak) – you see, he felt that *he* had trypophobia. He called this disorder 'the most common phobia you have never heard of'.

So how do you test for it?

Luckily (or unluckily) the head of a lotus has lots of fairly irregular small holes containing seeds. Just viewing a picture of a lotus head can set off trypophobia in a susceptible person. So, the authors asked 286 adults to look at a picture of a lotus head, and some 15% agreed that the image was 'uncomfortable or even repulsive to look at'. It turns out that looking at aerated chocolate, soap bubbles, frothy milk, strawberries, honeycombs, coral reefs and the like can all bring on the symptoms.

Why do some of us have this reaction to seeing holes?

Pronounce 'trypophobia'

The first syllable should probably be pronounced 'trip' (instead of 'tripe'), following the Greek heritage of the word.

33

DETOUR

Holey moley – smallpox was a big killer!

Think of all the human deaths that have occurred on Earth over the last 1,000 years. According to one estimate, smallpox caused 10% of all those deaths! Hooray for the vaccination that got rid of smallpox!

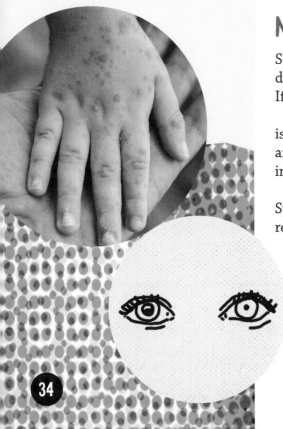

DANGER AND DISEASE

It turns out that certain highly poisonous animals have repeating patterns on their skin, scales or exoskeleton – often little dots or circles. These include the blue-ringed octopus, the Brazilian wandering spider (love that name!), the deathstalker scorpion, the inland taipan snake, the king cobra snake, the poison dart frog, the puffer fish, and the stone fish.

So, using a broad brush, it might make evolutionary sense for some of us to be a little scared of a critter that can kill.

Another study in 2018 pointed out that many virulent and infectious deadly diseases – such as smallpox, scarlet fever, leprosy, typhus and rubella – can cause clusters of pustules or roughly circular shapes (dots or spots) to appear on human skin.

Again, it makes sense to have the ability to instinctively identify and avoid other people who are suffering from a deadly infectious disease, especially back when there was no treatment.

So now we have two possible evolutionary reasons for developing a phobia of smallish, vaguely regular holes – first, to avoid poisonous animals and, second, to avoid sick, diseased people who might infect the healthy.

NOT A 'PHOBIA'?!

So now we're coming back to the Big Twist. Remember that the definition of a phobia quite clearly includes the concept of 'fear'. If there's no 'fear', then there's no 'phobia'.

But some research suggests that what happens in trypophobia is not 'fear', but 'disgust'. From a human reflex point of view, 'fear' and 'disgust' are very different. For one thing, fear makes the pupils in your eyes get bigger, but disgust makes them smaller.

If somebody suddenly attacks you, the Sympathetic Nervous System kicks into action, to prepare you for a fight or flight response. To get you revved up, your heart speeds up, the blood supply to your muscles increases – and your pupils get bigger (to let in more light, so that you can see better).

But if you see something disgusting, then a different pathway called the Parasympathetic Nervous System activates. In this case, your pupils get smaller.

And with trypophobia, you get smaller pupils!

So, first, it seems that 'trypophobia' might be a misnomer

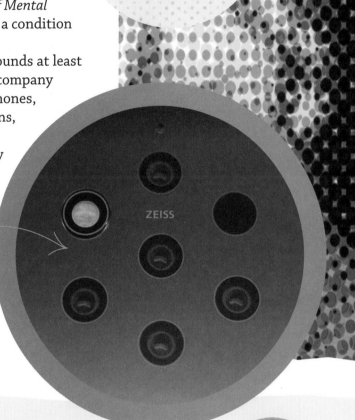

for this phenomenon – and that it's not really a phobia at all.

And, second, getting down to the nitty gritty, trypophobia is not recognised as a real disorder by the medical profession – at least not in 2019. (It's currently not listed in *DSM-5* – the fifth edition of the *Diagnostic and Statistical Manual of Mental Disorders*. *DSM-5* is the official 'bible' of psychiatry. If a condition is listed in *DSM-5*, it's considered legit.)

But let's accept that trypophobia is real (which sounds at least halfway reasonable). Early in 2019, the Nokia phone company decided to re-enter the phone market. One of their phones, the Nokia 9, had five cameras, each with a separate lens, plus a distance sensor and flash. And yes, the seven holes are arranged in such a way as to have apparently set off trypophobia (if you were the right distance from the phone) in some people. When you go to take a picture of someone with this phone and they see seven holes clustered on the back of it, some people will automatically look disgusted. Disgust is usually not your best 'face'.

So, now do we need software to automatically photoshop disgust into something more attractive? Is this just a flimsy new business model to sell us more apps, or should we stop taking so many photos?

TRYPOPHOBIA ON SOCIAL MEDIA AND IN TECHNOLOGY

Kendall Jenner (from the TV reality show *Keeping up with the Kardashians*) claims to be a trypophobe. She wrote, 'I can't even look at little holes. It gives me the worst anxiety. Who knows what's in there?'

In 2017, the seventh season of *American Horror Story* was advertised using images that included women with holes on their hands and faces. Apparently, this set off previously unknown trypophobia in so many people that the Twittersphere erupted with warnings and protests.

The tryphophobia support groups also warned fellow sufferers about the 2018 superhero movie, *Black Panther*. The character N'Jadaka (played by Michael B. Jordan) is the cousin of the hero, T'Challa. When N-Jadaka strips off his shirt, multiple rows of similar-sized scars on his chest (one for each person he has killed) are very obvious.

Disgusting clusters?

REST AREA

BIG BROTHER INTERNET?

George Orwell's novel *1984* describes the dystopian future of a world in a state of continual war, with oppressed citizens kept in line by continuous government surveillance wherever they go.

The omnipresent 'telescreens' not only broadcast propaganda from Big Brother and 'Fake News' 24 hours a day, but they also snoop on people with inbuilt video cameras and microphones.

Today, just like that fictional world, many of us have microphones in our homes. But the big difference is that the government didn't install them. No, we citizens voluntarily paid for them. These microphones are embedded in our smart devices, including our phones. They are controlled by software (digital assistants) such as Amazon's Alexa, Apple's Siri and Google Assistant.

Now here's two weird things about these devices.

First, a few of these digital assistants have been recording everything they hear. Now it's quite clear that they *listen* to everything you say. After all, they *have* to listen all the time, because they have to 'wake up' when you say special words such as 'Alexa' or 'Amazon', or 'Hey, Siri', or 'Okay, Google'. But *recording* everything you say, even when you haven't said the magic word, and when you are in a different room – that's quite different.

Second, their inbuilt microphones can be hacked. They can be taken over by sounds in your environment that you don't even notice. And then your smart device can do things you don't want it to do.

BUT IT'S SO CONVENIENT

Back in 2018, around 22% of Americans used voice recognition devices such as Amazon Echo and Google Home. These devices form part of the so-called Internet of Things (IoT) that many of us have joined. Our TVs, phones, toasters, garden taps, cars and even our shoes can all be linked to the Interwebs. It is definitely very convenient to be able to say, 'Hey, Siri, three-minute countdown' when you're boiling an egg. After all, you don't want to wipe your food-covered hands all over your nice, clean smartphone to start the

1984

George Orwell's book was published in 1949. It's been consistently listed in the Best 100 English-language novels. It gave us terms such as 'Big Brother', 'thoughtcrime', 'unperson' and 'doublethink'.

Nothing is what it seems in the world it describes. The main character, Winston Smith, works at the Ministry of Truth. His job is to rewrite historical records, to fit in with whatever the State deems correct. The State is run by four ministries, which have names that are the opposite of what they actually do (an example of 'doublethink'): 'The Ministry of Peace concerns itself with war, the Ministry of Truth with lies, the Ministry of Love with torture and the Ministry of Plenty with starvation.'

Maybe doublethink pervades our real world. Today, no country has a Ministry of War, only a Ministry of Defence. But if no country declares war, then why is there a need for any defence?

The Internet of Things

The term 'Internet of Things' (IoT) was probably first coined in 1999 by Kevin Ashton of Procter & Gamble. It refers to everyday objects (not just computers and smartphones) that are embedded with electronics and have some kind of connection to the Internet.

The IoT includes your fridge, your lighting fixtures, thermostats in your house, your rice cooker, your wristwatch, your toothbrush, and even the bell and lock for your front door.

These IoT devices can communicate and interact with other IoT devices through the Internet. They can be remotely monitored, or controlled.

According to Cisco Systems, the IoT came 'alive' somewhere between 2008 and 2009. That was the first time that objects ('things') connected to the Internet outnumbered people connected to the Internet.

The IoT is growing rapidly. From 2016 to 2017, there was a 31% increase in connected objects, bringing it up to a total of 8.4 billion. By 2020, there should be 30 billion IoT devices, with a total global market value of US$7.1 trillion.

timer. But some people are worried that these digital assistants are snooping on us when we don't expect them to.

Both Google and Amazon claim that their devices record what they hear *only* after they have been deliberately triggered by a specific voice command. (Some devices have a physical switch that can turn the microphone off.)

'SNOOPING' BY GOOGLE

But that hasn't always been how it's worked out in real life.

In 2017, a tech journalist found that his Google Home Mini was recording him almost 24/7, thanks to a glitch in the device. In fact, this had happened to all the other journalists to whom Minis had been distributed.

More recently, Google filed a patent application targeted at children. This proposed device would sample a child's speech patterns, whispers or silence to see if he or she was engaging in 'mischief'. It could work out if the child was moving, by listening for sounds linked to specific actions, such as dragging a kitchen chair across the floor and then climbing onto it. The smart speaker could then 'provide a verbal warning', such as 'Don't eat the chocolate'. Sounds a bit like a baby monitor on steroids.

This Google patent application also discusses 'snooping' using the camera in your device. It might, for example, see Will Smith's face on 'a T-shirt on a floor of the user's closet'.

Would it tell you to pick it up and stop being messy? No.

Instead, it would search your web browser history. If you had recently searched for 'Will Smith', it would then 'provide a movie recommendation that displays, "You seem to like Will Smith. His new movie is playing in a theatre near you"'.

'SNOOPING' BY AMAZON

In another case of digital assistants recording everything, Arkansas police were investigating a possible murder. (A person was found dead, face down in a hot tub, after a night of heavy drinking and watching football with friends.) The defendant gave the police permission to retrieve any relevant recordings from Amazon. The police got the voice recordings captured by an Amazon Echo – a smart speaker with Alexa – on that day in November 2015. The defendant was acquitted.

PROBLEMS WITH THE INTERNET OF THINGS

One problem with the Internet of Things (IoT) is what Jean-Louis Gassée (one of Apple's initial alumni team) calls the 'basket of remotes'. You've probably seen someone's home entertainment system controlled by half a dozen remote controls, none of which can control more than one device. So he envisages a situation in which we have thousands of devices that are run by hundreds of different applications – and that don't have common protocols for communicating with one another. Each company wants to keep to the protocol it invented, to avoid retraining engineers and to keep the buyer locked into buying their related products (e.g. Canon lenses don't fit on Nikon camera bodies,

and vice versa). This is an example of how business practice can interfere with progress.

Loss of privacy is another problem. People's behaviour and conversations in their own homes are being spied on by their televisions, thermostats and even kitchen appliances.

Security in the Internet of Things is poor. The IoT developed so rapidly that there was no proper consideration of security challenges. For example, some cars communicate the pressure in each tyre to the car's central computer, via Bluetooth. Now Bluetooth has a range of only several metres. However, that's enough for one vehicle to 'attack' another on a multi-lane highway. Hackers in one car successfully have attacked another car's central computer – via the unsecured Bluetooth.

What could they control? Lots. Today's cars have computers controlling the internal lights, external lights, the brakes, the engine, the door locks, the horn – in other words, practically everything. In 2008, as a laboratory exercise, hackers were able to remotely control a pacemaker in a person's chest. They could also control implanted cardiac defibrillators, as well as insulin pumps.

Another problem is service (or repair) of the devices that you bought. Many people purchased home automation devices from a company called Revolv, partly attracted by the promise of a Lifetime Subscription. Another company took over Revolv, and shut down that Lifetime Subscription. But interestingly, that other company was not a small, poorly funded, fly-by-night company. It was controlled by Google, one of the wealthiest tech companies in the world.

That's the good news (for the defendant). The bad news is these recordings included a conversation about football, carried out in a separate room. This conversation was *not* triggered by an Alexa command. So, that tells us that back in 2015, the Amazon Echo was recording, and storing, every single conversation that it heard in the defendant's house. Maybe it was doing this in your house too?

Delving into the patents of companies like Amazon and Google is a fascinating exercise into the future of privacy. In one of its patents, Amazon describes what it calls a 'Voice Sniffer Algorithm'. This would let the Amazon digital assistant listen to a phone call, and monitor the conversation for words such as 'like', 'bought' or 'dislike'.

So let's suppose that you had mentioned (in the privacy of your home) that you hadn't been to the movies for a long time. Well, the very next time that you accessed your smartphone or computer to check the weather, you might get a personalised advertisement for the local movie theatre. Different personalised ads could pop up if you had talked about being pregnant, or liking indoor climbing.

(We have now been conditioned to expect that our searches on the web are no longer private. Search for something on the web, and ads related to the search term will pop up for weeks to come. If you contact your friends about an item using Google Mail or Facebook Messenger, it's very likely that you'll be sent ads for that very item. Unfortunately, we have come to expect that as the new 'normal'. But that's different from *talking* to a friend about it, either over the phone or in your house, face to face.)

On the one hand, Amazon says it takes privacy seriously, and would 'not use customers' voice recordings for targeted advertising'. But on the other hand, the patent application deals with the exact opposite. (Nice doublespeak?) The patent includes a diagram showing the digital assistant listening to a phone call between two people, specifically scanning for words with commercial possibilities. Further down the line, those people could receive two quite different, targeted ads – one for wine, the other for a zoo visit.

The vacation was wonderful. I really enjoyed Orange County and the beaches. And the kids loved the San Diego Zoo.

Identified person
(Laura)

When we went to Southern California, I fell in love with Santa Barbara. There were so many great wineries to visit.

Verified user

Identified keywords	Laura	Verified user:
	• Orange County	• Santa Barbara
	• beach	• winery
	• San Diego Zoo	• wine
	• zoo (kids)	
	• animals (kids)	

The non-profit advocacy group Consumer Watchdog studied many of these patent applications and summarised: 'It's really clear that this is spyware and a surveillance system meant to serve you up to advertisers.'

OTHERS CAN TALK TO YOUR DEVICE

Over the last decade, many of us have got used to the idea that our smart devices will listen to us. And today it seems quite normal that we talk to them to set an alarm or play some music.

But isn't this supposed to be a 'special' relationship ... that your digital assistant will obey *only* your command?

Sorry to break your heart, but *no*.

Unauthorised people can secretly talk to our various digital devices. They can get our devices to do things that we don't want them to do – such as buy stuff. And all three of the major digital assistants (Siri, Alexa and Google Assistant) have been hacked.

Ongoing research found that your devices can be controlled remotely in three different ways.

First, unnoticed background white noise could control your device. Second, voices (such as from a podcast, a song or a TV) could control your device. Finally, sounds you couldn't even hear could control your device!

CONTROLLING YOUR DEVICES REMOTELY

In 2016, students from Georgetown University and the University of California were able to hide surreptitious commands in white noise that you would normally ignore. This white noise came either through your sound system and loudspeakers in your house, or in YouTube videos from your smart TV or computer.

These commands were surprisingly powerful. They could make your smartphone go into aeroplane mode, or even open a website. Both Amazon and Apple are installing protections to stop this.

By 2017, the students from the University of California had gone one step further. They successfully embedded commands, not into white noise, but into spoken voice (such as a podcast) and even recordings of music. Again, these commands were completely unnoticed by the owner of the smartphone or device. And again, the smartphone or device was successfully hacked.

The net was originally designed by the military to be a robust communications system to launch nuclear weapons.

Who knew that the same thing would be used nowadays to snoop on you?

The first smartphone: $40,000 per GB!

Back in 2009, our family walked some 790 km across Spain on El Camino de Santiago. I carried the first really functional smartphone, the iPhone 3G.

It was amazing, compared with the previous mobile phones I'd used. Besides phone calls and texts, I could take photos, have Internet access to websites and email, translate between Spanish and English, write notes, record my voice, play music, use a calculator and even navigate on maps.

Today's smartphones even have magnetometers, accelerometers, barometers, gyroscopes, proximity sensors, fingerprint sensors, facial recognition and ambient light sensors.

Each day for five weeks during this walk, I sent photos back to my ABC website in Australia. Data was expensive back then, and especially much more expensive for short-term foreign travellers who didn't have access to a local data plan. In 2009, as a foreigner in Spain, the data cost me $40 per MB, which worked out to be $40,000 per GB! (Luckily, I didn't send anything close to a single GB of data.) Each photo was about one-third of a MB, which worked out to be a total of $40 per day.

Finally, again in 2017, another group of researchers, from Zhejiang University in China and Princeton University in the USA, were able to activate a device's voice recognition systems using high-frequency sounds that the human ear cannot hear. The researchers called this technique Dolphin Attack (probably because dolphins use high frequencies we cannot hear with our ears). So you wouldn't even hear a hiss of white noise, or a voice, or music. As far as you were concerned, the environment was silent. But again, your device was taken over – in this case, by ultrasound.

And what did these hacked smart devices do? They made phone calls, took pictures, sent texts and visited 'dangerous' websites. The hackers even got into smart home accessories such as your front door lock. Somebody could walk up to your smart front door, play it an inaudible track from their phone, and your door would open. This sort of hacking is probably the scariest of the lot.

The possibilities for cybercrime are enormous.

THE FUTURE?

By the year 2021 in the USA, humans should be outnumbered by smartphones and smart speakers with digital assistants. At the moment, Apple's Siri, Google Assistant and Amazon's Alexa can all be tricked into obeying commands that we humans cannot detect, and which can be surreptitiously embedded into voice or music.

Mind you, the companies will fix that specific fault – but the baddies won't stop. They will invent something new, and the battle will continue.

Google and Amazon can let you read the transcriptions of what you have said, and let you hear the actual recordings of what you said.

The technology is moving so quickly that it's difficult for the average non-techie consumer to protect themselves. And so the tech companies (who do understand the technology) have a moral responsibility to balance security against ease-of-use by the consumer.

In terms of privacy, perhaps it's time for a return to sign language and invisible ink, or perhaps telepathy will be our saviour?

I really hope in the future I won't be saying – 'Big Brother, set the egg timer for three minutes.'

RUNNING OUT OF INTERNET ADDRESSES

The Internet has been around since the 1980s. The World Wide Web started around Christmas 1990. Every device that connects to the Internet needs an IP (Internet Protocol) address. Your router/modem at home has one. Add another IP address for your smartphone, and you're up to two. And that's just for one person out of the 7.5 billion on the planet.

So it's not surprising that we started running out of IP addresses around 2011. Yep, it's not only first-time homebuyers that struggle with the housing market – even electrons couldn't have their own address!

The first system was IPv4 (Internet Protocol version 4). It allowed for 2^{32} (4,294,967,296 or just over 4 billion) addresses. That's fewer than one per person on Earth. It ran out in early 2011. The lifetime of IPv4 has been extended to some extent by 'reusing' IP addresses in Private Networks.

Fortunately, Internet folk had seen what was coming, and had devised IPv6. (I don't know what happened to IPv1–3, or 5.) Major computer operating systems since Windows 2000, and Mac OS 10.3 are compatible with IPv6. IPv6 and IPv4 co-exist at the moment, but we are slowly shifting over to pure IPv6.

IPv6 provides 2^{128} or 340 trillion trillion trillion separate addresses. That's about 50,000 trillion trillion addresses for each of us. Or if you want to pick a really obscure measure, if each person on Earth had a tonne of pure carbon, then each atom in that tonne could have its own unique IP address. We can, however, certainly have as many IoT devices as we want – but didn't they say that with IPv4?

CORAL SPAWNING

There are two things most of us 'know' about the Great Barrier Reef. First, we 'know' it's big. And second, we 'know' that coral spawns once every year.

Most people expect that the coral on the entire reef do their spawning thing in a giant sexual paroxysm, synchronised to the very second all the way up and down the reef. No wonder the tabloids write headlines like 'It's the Largest Orgasm on the Planet'.

Well, yes, the Great Barrier Reef is enormous. But, no, coral spawning doesn't happen everywhere at the same time. (By the way, 'spawn' can be eggs, or sperm, or both jumbled together, or fertilised eggs.)

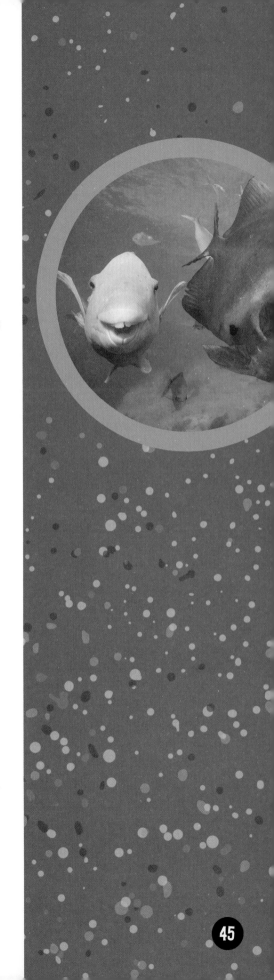

GREAT BARRIER REEF

The Great Barrier Reef, off the coast of Queensland, has a surface area about one and a half times bigger than the state of Victoria – about 350,000 km². (That's roughly one-thousandth of the combined area of all the oceans on Earth.) It's about 2,300 km long and up to 250 km wide. It has over 2,900 individual reefs and some 900 islands. More than 1,600 species of fish, 133 species of sharks and stingrays, and over 30 species of whales and dolphins have made it their home.

It's also the biggest single structure built by living creatures. And you can see it from space, unlike the Great Wall of China. However, it's not as 'Great' as it used to be. Global Warming/Climate Change has been tough on it.

The first worldwide coral bleaching event was in 1998, and 70% of the world's coral reefs were affected. That was the first time that the Great Barrier Reef was bleached on a large scale. There had been smaller regional bleaching events in 1982–83, in a few locations around the world. We've known that when coral is 'unhappy' it turns white – e.g. due to local flooding, sudden cold weather and widespread local heating.

Today, carbon dioxide levels are soaring above 415 ppm (parts per million), and so major coral bleaching events have been hitting the Great Barrier Reef even harder, and more frequently.

In the 27 years between 1985 and 2012, half of the cover (the area of the reef occupied by corals) on the Great Barrier Reef disappeared. If you extrapolate that trend, half of the remaining coral animals will be gone by 2022. But worryingly, the trend towards destruction is increasing and getting worse …

CORAL 101

Coral reefs cover roughly 0.1% of the area of the world's ocean, but they support over 25% of all ocean creatures.

The Great Barrier Reef was made entirely by trillions of stationary animals called coral polyps, which are connected together to make coral colonies. (They're animals that do not move around, but instead are cemented down in one location.)

Today, there are over 400 known different species of corals inhabiting, and creating, the Great Barrier Reef. Their work in building and maintaining the complex 3D structure of the reef makes it possible for many thousands of other totally different

Ocean pollution and rising egg-sperm bundles

A newly understood threat to coral spawning is dirty ocean water. This dirt can come from dredging, dumping soil or sludge near a coral reef, runoff of land pollutants via a river, and even natural re-suspension events.

The grains of sediment can stick to the buoyant egg–sperm bundles while they are trying to rise to the surface. In some cases, these grains of dirt can weigh down the egg–sperm bundles, and stop them from getting to the surface of the ocean.

The numbers of egg–sperm bundles that get to the surface can be reduced by 10–15%, depending on the depth of the coral, and the level of pollution.

species to come together. In sweet, intricate combination, they survive and thrive.

Coral polyps are shaped like cylinders – usually a few millimetres wide, and a few centimetres long. One end sits on the coral's skeleton. At the other end is their mouth, surrounded by tentacles. Food comes in and leaves through their mouth. So, yes, coral are animals that eat with their bum – or, more politely, their excretory organ.

CORALS GO BACK A LONG WAY

The first animals to have coral-like hard shells were *Cloudina* – tiny filter-feeding animals. They appeared around 550 million years ago. Within a mere 2 million years, they were living in the first known coral reef – about 300 m thick, and about 7 km long.

So how do corals make babies, if they're stuck in one spot? Well, let me tell you, coral sex is mind-boggling.

CORAL SEX 101

First, let's look at the 'sex patterns' of these strange animals.

Averaged across the world, about 71% of coral species are genuine hermaphrodites. This means that each polyp has both male and female bits. But about 26% of coral species are *either* male or female. They are one *or* the other – but not both at the same time. In each colony, they are all of one sex. That leaves about 3%, that either have mixed sex patterns, and/or they can swap sexes during their lives.

Second, let's look at how they make 'babies' (or coral larvae).

Coral usually dump their eggs and sperm into the surrounding water. An egg and sperm get together and create a fertilised egg. This is called 'external fertilisation' and is the most popular way to make coral babies.

About 65% of coral babies come from hermaphrodites that broadcast their eggs and sperm into the water. The eggs get fertilised, and the larvae 'settle down' about four to seven days later.

About 20% of coral babies come from corals that are either male or female, which broadcast their eggs or sperm separately into the surrounding waters.

But some coral animals will go down a different pathway. They release sperm which internally fertilise their own eggs, and then the

coral 'broods' the developing larvae. When ready, they will release fully competent larvae, which can start up a colony of their own.

Then about 15% of coral babies come from brooding species – these can be hermaphrodites or they can have separate sexes. These larvae tend to settle down nearby within a day.

The remaining tiny percentage of babies come from coral species that have mixed sex patterns, or swap sexes.

I told you the sex was complicated. But from here on, everything else is easy.

CORAL SPAWNING

So what about the legendary coral spawning? Does it really happen all at once, in an instant, across the whole reef?

Well, some individual corals do all their spawning in a single second – that's right, an explosive burst in just one second. But others do it for minutes, or hours, or in shifts. So it's definitely not an instantaneous event for all coral.

On the Great Barrier Reef, the timing of coral spawning varies. For the inner reef (closer to shore), which gets warmer earlier, coral spawning happens during the week after the full moon in October. But for the outer reef (further out to sea) and southern parts of the reef, which have colder waters, coral spawning happens after the full moon in November–December. I've been lucky enough to see coral spawning at Heron Island, on the southern Great Barrier Reef.

On average, the coral release their eggs and sperm sometime between day 3 and day 7 after the full moon. There are many factors involved in this timing, including temperature of the water, intensity of the moonlight, and more.

So the coral do *not* all spawn in the same month – much less the same week, day, or hour or second. Corals that brood their larvae typically release them on a monthly cycle, during multiple summer months.

SPAWN MAKES BABIES

The timing of spawning is important. About a week after the full moon, the difference between high and low tides is smallest. This makes it easier for the eggs and sperm to 'get together' and not get washed away.

DR KARL'S Q+A

The Moon and humans

Does the Moon make a difference to coral having sex? Yes.

But what about this idea that the full moon makes the lives of people like police officers and hospital doctors busier?

Yes, but only if you believe hearsay such as, 'It was really busy last night in the Emergency Department – there was a full moon, so no wonder!'.

But the real answer is no, according to studies that look at time periods covering years and hundreds of thousands of people.

So what about Shakespeare's *A Midsummer Night's Dream* and the 1987 movie *Moonstruck,* and all those werewolf movies? The full moon plays a major part.

Ah, that's called Artistic Licence.

DETOUR

Eyes and clocks in coral?

Yes, it makes lots of sense for the coral in an area to all release their sperm and eggs around the same time so they can pair up. But they don't have calendars, clocks or eyes – how do they know when to do it?

The answer is 'circadian rhythms' – the Rhythm of Life. Many creatures have internal clocks that respond to changes in light and temperature.

We have found 'eyes' in at least one type of coral. They are sets of chemicals ('cryptochromes' in the body of the coral, and made by the coral) that respond to light levels, and then set off chemical reactions. They are not eyes like the ones we have, but they respond to light. So they respond differently, depending on whether it's a full moon or a new moon.

It seems that coral's own internal clocks get them primed, and that the Moon gives the final trigger for their choreographed spawning.

Sometimes spawning is so bountiful that it looks like an underwater snowstorm, with tiny white and pink flakes drifting upwards.

In the case of some brain coral, each 'spawn' is a blob about half a millimetre across, and pinkish in colour. Each blob contains one or more eggs, but also many, many thousands of sperm.

The blob rises slowly to the surface, and then pops open, allowing the eggs and thousands of sperm to separate.

Usually, the sperm will not fertilise an egg with the same genetics, from the same coral parent. This allows future coral to have greater genetic diversity, and so, greater resilience. But after a few hours, the sperm run the risk of being wasted – so if they're running out of time, they will mate with a genetically related egg nearby. It's a coral case of 'If you can't be with the one you love, then love the one you're with'.

At one time, it was assumed that coral gave birth to fully competent larvae. But in 1981, marine biologist Peter Harrison and an informal team of friends, fellow students and colleagues spent months camping out on the reef, trying to document what was really going on. They were astonished to see a massive underwater blizzard of pinkish coral spawning – made of bundles of eggs and sperm, not little coral babies. Harrison had the largest set of data, so he became the lead author on that first famous paper on coral spawning that was published in *Science*, back in 1984, introducing the concept of the mass synchronised spawning.

SETTLE DOWN

Once fertilised, the soft-bodied larvae float around. They are attracted to light (so they head for the surface, rather than the darker depths) and sound (so they head for waves crashing onto a reef, rather than open waters). They have been known to float with ocean currents for hundreds of kilometres before descending to start a new colony.

If lucky, they will land on something hard, and 'absorb' microscopic algae into their bodies. Algae are coral's essential symbiotic partners in a remarkable long-term relationship. The time between spawning and settling down is usually four to seven days, but occasionally can reach two months.

If you get enough of them in one place, you end up with a reef. In a sense, coral is a 'living rock'.

APRIL 2019 ...

Unfortunately, the increasing number of coral bleaching events means fewer breeding colonies. In turn, this means less coral spawning.

Marine biologist Professor Terry Hughes reported in April 2019 that the numbers of new baby corals had fallen by as much as 95% in some parts of the northern (and therefore hotter) parts of the Great Barrier Reef.

This massive drop was a direct result of totally unprecedented back-to-back mass coral bleaching events in the previous two summers. Professor Hughes told the *Sydney Morning Herald* that the reef 'wasn't losing the capacity to produce babies. There just weren't enough adults'.

The hermaphrodite corals that spawn eggs and sperm suffered the worst. Their spawn numbers dropped by up to 98%. The corals that do 'internal brooding' and release fertilised eggs didn't do as badly. Their numbers dropped by about a third. In fact, thanks to this massive change, internally fertilised brooding larva outnumbered externally fertilised larva for the first time. This is an unprecedented change in behaviour for coral.

Unfortunately, it seems that reefs need a longer time than we thought to recover from bleaching events. Professor Hughes thinks that at least five years, and more probably ten years, have to pass before the adult population returns to pre-bleaching levels after a major bleaching event. This is a nasty surprise.

The Great Barrier Reef has bleached once every five years in the last two decades. With carbon dioxide levels rising, this can only get worse.

Professor Hughes wrote, 'It will take at least a decade for the fastest growing species to recover, and much longer for longer-lived and slow-growing species.'

It may be that our grandchildren won't get to see the Great Barrier Reef in all its blazing, kaleidoscopic glory and with its vast biodiversity.

What will become of Nemo the clownfish if he has no home?

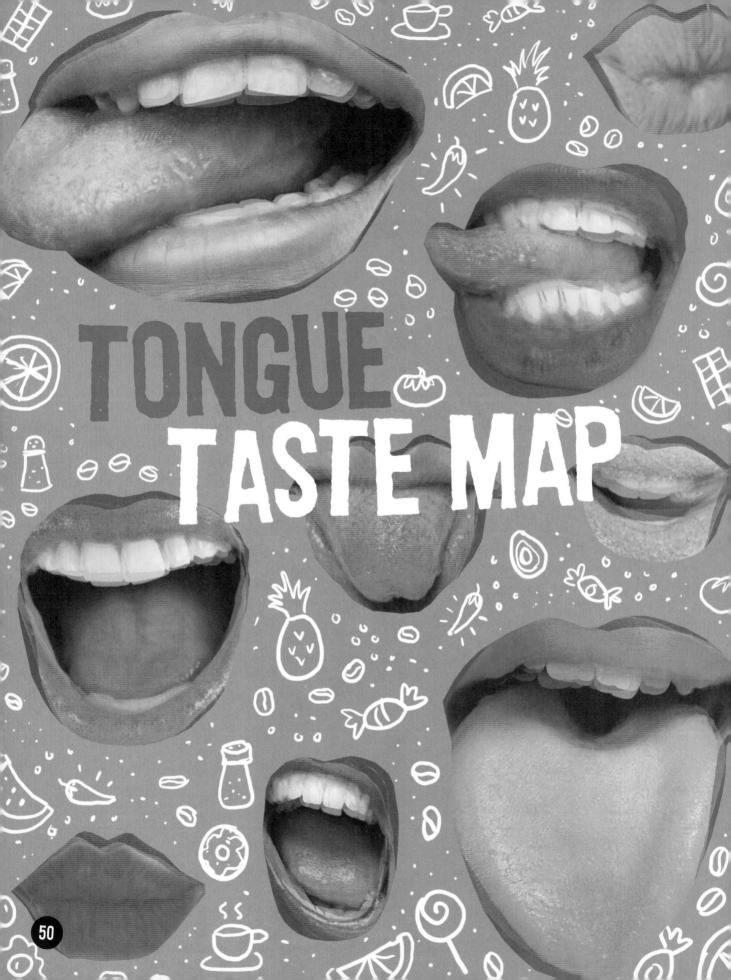

TONGUE TASTE MAP

How does your brain know if you're eating chocolate or broccoli? The answer is 'taste'. But even today, taste is still the most poorly understood of the supposed five senses of Touch, Sight, Smell, Hearing and Taste.

For example, consider the famous Tongue Taste Map (see p. 54). It shows four separate tastes, each located in a different part of the tongue. Spoiler alert – it's wrong.

TASTE 101

Let's start by looking at the history of Taste.

About 2,500 years ago, in the 4th century BC, the Greek thinker Democritus came up with one of the first theories to try to explain taste. Democritus thought that sweet substances were 'round and large in their atoms', while sour substances were 'large ... rough, angular and not spherical'. Bitter substances were 'spherical, smooth, scalene and small', while saltiness was caused by 'isosceles' atoms. Surprisingly, 2,500 years ago, he had uncovered a big chunk of the truth – different tastes are related to different shapes of chemicals.

Taste is about more than just food preferences – taste keeps us alive. It tells us what's nutritious, and also warns us of potential toxins.

So, the sensation 'sweet' tells you that lots of calories/kilojoules are here, while 'bitter' could warn you of potential toxins. The sensation of 'salt' could alert you to the presence of essential electrolytes, while 'sour' might warn you against spoiled or unripe fruit.

Taste also triggers other reactions. The taste of peppermint increases the production of saliva in the mouth, while the taste of cinnamon increases peristalsis in the gut.

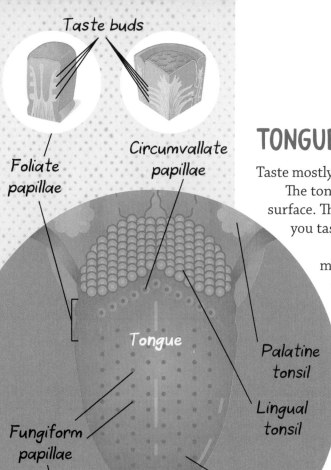

Taste buds

Foliate papillae

Circumvallate papillae

Tongue

Palatine tonsil

Lingual tonsil

Fungiform papillae

Filiform papillae

TONGUE 101

Taste mostly, but not always, happens on the tongue.

The tongue is not smooth – it has quite a bumpy and velvety surface. These 'bumps' are called papillae. They are crucial to how you taste foods.

There are four different types or shapes of papillae. The most numerous type is the filiform papillae. They are all over the tongue – but they have nothing to do with taste. They do not contain taste buds. Instead, they are related to feeling mechanical pressure or sensation.

The other three types of papillae are fungiform (meaning 'like mushrooms', located at the front of the tongue), circumvallate (meaning 'like a wall', at the back), and foliate (meaning 'like a leaf', on both sides, towards the back of the tongue). These each contain hundreds of tiny taste buds. So they are connected to the sensation of taste.

TASTE BUDS 101

The average human tongue has between 2,000 and 8,000 taste buds (but it can vary between 500 and 20,000). On average, a taste bud will survive for about seven to ten days before it dies and is replaced.

A taste bud looks like a tiny sphere or onion. Taste buds themselves are in turn made up of taste receptor cells. These cells are clustered together, within each taste bud, in groups of about 50–100. As a group, they look like the segments of a kind of mini orange.

The taste buds lie between the cells that make up the surface of the tongue. Each one has a tiny hole (or pore) opening onto the surface. The taste receptor cells extrude very fine hair-like filaments (microvilli) upwards through this tiny hole, and into the saliva that coats the tongue. These filaments 'sense' or monitor the various chemicals that come and go on the surface of the tongue. At the bottom end of the taste bud is a bundle of nerves that connect to each taste receptor cell. These nerves carry the taste sensation towards the brain to be processed.

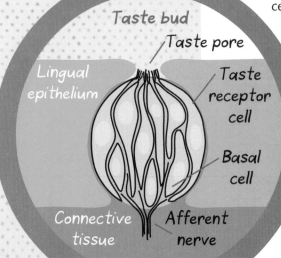

Taste bud

Taste pore

Lingual epithelium

Taste receptor cell

Basal cell

Connective tissue

Afferent nerve

SUPERTASTERS

There is a huge variation from one human to the next, in how well we detect some tastes. Some of this difference is genetic, while some can be environmental. (For example, when babies grow into adults, they tend to prefer the food that their mother ate while they were in her uterus.)

One chemical used to test this genetic difference is called PROP (6-n-propylthiouracil). Some people (called 'supertasters') perceive PROP as intensely bitter. However, 'medium tasters' recognise it as only slightly bitter, while 'non-tasters' cannot taste it at all. This sensitivity is related to the number of taste buds on the tongue, and their internal differences.

Supertasters have fungiform papillae that are a lot smaller in size than most people's, but they have a lot more of them. Their fungiform papillae are about half the area of those of non-tasters – 0.38 mm^2 versus 0.67 mm^2. Supertasters have 425 taste buds per sq cm, medium tasters have about half as many (185), while non-tasters have only a quarter as many (96 taste buds per cm^2).

Mind you, the so-called 'non-tasters' are non-tasters only for the test chemical, PROP. They can still register the sensation of bitterness for other chemicals, such as quinine.

After all, being able to taste 'bitter' is so important to survival that the average person has over two dozen different types of receptors to detect different variations of 'bitter'.

And, as an interesting aside, non-tasters are more likely to be alcoholics. Why? Another thing we don't know.

The main factor involved in how you taste stuff is genetic – in other words, inherited.

TASTE AND EMOTIONS

Your emotional state can very strongly influence how food tastes.

For example, when people are severely depressed, their sensitivity to different tastes is much reduced. However, this sensitivity returns to normal when they recover. This altered taste has long been recognised as one of the 'neglected symptoms of depression'.

We know that blood levels of serotonin and noradrenaline are altered in anxiety and depression.

In one study, the scientists used drugs that specifically

Conditioned taste

Working in the book trade means working with people who read books all the time – and I love that. Even better, they like to party and let their hair down.

But taste can change, as has been demonstrated by some of my friends, who inadvertently became averse to champagne, a drink they once enjoyed!

They went to a Christmas party where champagne and seafood were served to guests. They happily washed down their seafood with champagne. But the seafood had gone bad – and they started vomiting.

The good news is that they rapidly recovered. The bad news is that they got conditioned, by association, against the champagne. They can't drink champagne anymore. Even the smell of champagne makes them feel like vomiting.

How tough!

DR KARL'S
Q+A

Acquired tastes

How come we each have different 'tastes' that we like and dislike, and how do they change with time?

First, there's some genetics involved. Some of us are wired in our DNA to like, or dislike, bitter tastes.

Second, as we get older, these preferences sometimes change. Avoiding bitter tastes is good for children, because many poisons are bitter. But liking bitter tastes can be good for adults, because some bitter chemicals (e.g. sinigrin, in broccoli) have mild anti-cancer properties.

Third, we can get 'conditioned' into liking (or disliking) a taste. So you might not like the bitter taste of coffee or alcohol, but you do like how it makes you feel – so you begin to 'develop a taste' for something you naturally dislike.

increased the levels of serotonin. This made the subjects more sensitive to sugar and quinine.

Other drugs that increased noradrenaline levels made the subjects more sensitive to the tastes of bitter and sour.

There is a biochemical basis to this. It turns out that both serotonin and noradrenaline directly affect the excitability (or sensitivity) of taste cells by changing how the 'ion channels' operate.

People suffering from an episode of panic disorder have a reduced sensitivity to bitter tastes. On the other hand, people who are under stress become more sensitive to the bitter aftertaste of saccharin.

TASTE AND OBESITY

Obesity also affects how you taste what you eat. Many studies show that in rats and in humans, being obese makes them less sensitive to the taste of 'sweet' – and makes them crave sweet foods even more.

With modern processed foods, the ever-increasing amounts of sugar and fat overstimulate the taste centres in the brain. Over time, this makes these centres less sensitive.

With decreased sensitivity, we look for more 'deliciousness' in our foods – e.g. by adding more sugar to our tea or coffee.

TONGUE TASTE MAP MYTH

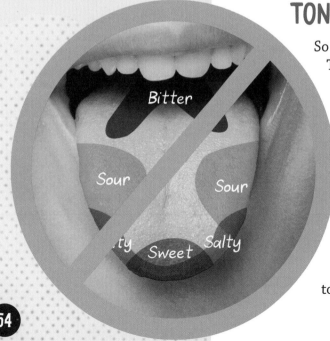

So now we are ready to tackle the Myth of the Tongue Taste Map …

Most of us have heard that there are four basic taste sensations, and that they are each localised to a specific part of the tongue. You might have seen the Tongue Taste Map showing that we sense sweet at the tip of the tongue, sour along the sides, and bitter at the back of the tongue.

This map is wrong in two different ways. First, there are (at least) five basic taste sensations not four. Second, we detect all five taste sensations all around the tongue.

How did this wonky Taste Map get passed along to generations of school and university students?

PART 1 - 1901

It began way back in 1901, when David P. Hänig, published his PhD thesis in the journal *Philosophische Studien*. He wanted to test just where the taste buds were most densely distributed. He also wanted to see if the then-known four basic taste sensations were detected differently across the tongue.

He used sucrose to test for the sensation of 'sweet', and quinine to test for the sensation of 'bitter'. He also used very dilute hydrochloric acid to test for the sensation of 'sour' and, of course, used salt to test for 'saltiness'. Dr Hänig found that all four tastes could be detected around the whole perimeter of the tongue. Yes, there were differences around the perimeter of the tongue, but they were quite small.

For example, 'sweet' was best detected at the tip of the tongue and worst detected at the base of the tongue – but it was still very well detected all around the perimeter of the tongue.

The sensation of 'bitter' was the opposite – worst at the tip, best at the base, but still very well detected all around the perimeter of the tongue.

And, similarly, 'sour' was best detected at the middle of the sides of the tongue, and worst detected at the tip of the tongue – but it was still picked up all around the edge of the tongue.

Finally, the sensation of 'salt' was tasted fairly equally all around the outside of the tongue.

Dr Hänig then made a rather impressionistic and very rough graph of his findings. It included only sweet, sour and bitter – not salt. It was such a crude graph that it didn't even have any values or numbers on the vertical axis (and that's a big no-no in science). He just wanted to show that for the sensations of sweet, sour and bitter, there was a slight variation around the perimeter of the tongue.

Dr Hänig's original graph from 1901

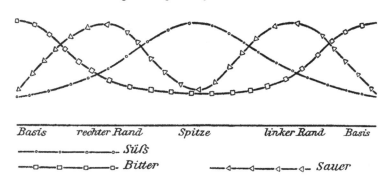

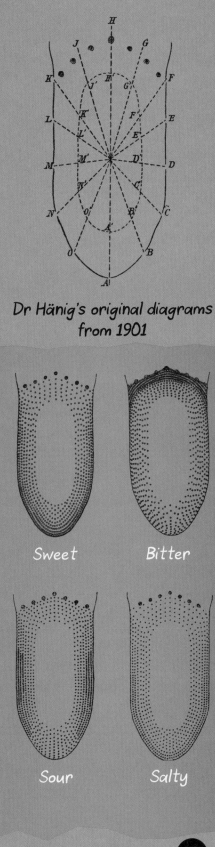

Dr Hänig's original diagrams from 1901

Sweet   Bitter

Sour   Salty

PART 2 - 1942

The next stage in manufacturing this tongue map myth took four decades to happen.

In 1942, the distinguished experimental psychologist Edwin G. Boring wrote an enormous 660-page summary of everything that was currently known about experimental psychology.

His book was a monumental effort. He mentioned Dr Hänig's name just once, on page 452. Someone (we don't know who) turned Dr Hänig's unlabelled graph into a new graph with labels, and linear scale markings on both axes. The sensation of 'salt' was arbitrarily added to Dr Hänig's original graph. The shapes of the curves were 'massaged'. And a few extra mistakes were made in translating from German to English.

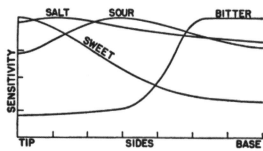

Boring's quite different graph from 1942

At no stage did Dr Boring claim that there were only four tastes (sweet, sour, salt, bitter). In fact, he specifically mentioned that other earlier workers in the field of taste (going back to the 1500s!) had mentioned up to 16 possible tastes.

Neither did Dr Boring say that the sensation of 'sweet' was greatest at the tip of the tongue and virtually zero everywhere else.

PART 3 - UNKNOWN

But other unnamed people did misinterpret Dr Boring's version of Dr Hänig's original work. They made the familiar Tongue Taste Map, with the sensitivity for sweet only at the front, sour only at the sides and bitter only at the back. And that's how the Tongue Taste Map came into existence, to be taught to generations of us.

But the map was so easy to test and find wanting. How could people ignore reality for so long? This map claims that sweetness, and only sweetness, is detected at the tip of the tongue. But if you place some salt on the tip of your tongue, you definitely taste salt.

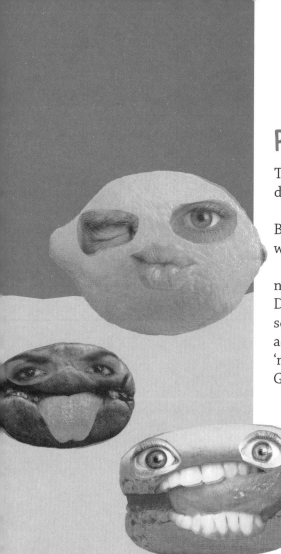

Taste is weird

'Taste' is a good example of 'evolution does not have to be perfect, just good enough'.

We humans cannot digest starch directly, and we can't taste it, either. Fair enough. And cats can't taste 'sweet'. Fair enough, they eat meat, not lollies.

But rats cannot digest starch, yet they can taste it. How come?

PART 4 – 1974

In 1974, another scientist, Virginia Collings, tried to set the record straight. Indeed, she was one of the first to challenge the Taste Map, which had been taught to one and all since 1942 (and is still repeated, even today!).

She started by trying to replicate David Hänig's original work. She did find slight (almost insignificant) variations in basic tastes around the tongue. This fitted in with what Dr Hänig reported in 1901. However, she disagreed with a few of his findings, especially with regard to the taste of 'bitter'.

She also found taste buds in places other than the tongue – including on the soft palate on the roof of your mouth, and on the epiglottis (the movable flap that stops food going down your windpipe, also called the trachea).

But there's another really obvious thing wrong about the four basic tastes being mapped onto the tongue.

There are at least *five* basic tastes. The fifth one is called 'umami'.

Who knows, in the future maybe we'll identify more than the five basic tastes. Maybe for that, we'll need the sixth sense …

UMAMI IN UTERO

Umami as a taste was first proposed by the Japanese scientist Kikunae Ikeda in the early 1900s. It is essentially the taste of the amino acid L-glutamate and a few 5'-ribonucleotides (e.g. guanosine monophosphate, inosine monophosphate). (These are chemicals that can be found in proteins and RNA, but not in fat or carbohydrates.) The chemical MSG (monosodium glutamate) has a very intense umami flavour.

Before we are born, we actually float in a clear umami soup – amniotic fluid. And the umami diet continues after birth. Breast milk has ten times more glutamate than cows' milk.

Glutamates have been prized in cooking for thousands of years. They are at very high levels in fermented fish sauces (ancient Rome, China), soy sauce (China) and fermented barley sauces (medieval Byzantine and Arab cuisines). Parmigiano cheese is a veritable umami bomb, as is the sun-dried tomato. Umami is also found in some Japanese sea vegetables, meat broths, grains, beans and well-grilled meats. It's also at high levels in Vegemite and Marmite.

REST AREA

BALLOON POPPING BAAM!

You know the party is really rocking when the party balloons are popping.

And the next morning, you can turn the clean-up into a science experiment by checking out the remains of the popped balloons. To your surprise, you notice two quite distinct patterns: some balloons are shredded into dozens of tiny slivers, but the rest end up as just a few larger fragments.

Welcome to the land of 'fragmentation'. This is where big lumps of stuff fragment (or break) into smaller pieces. Fragmentation happens everywhere. It happens in space when asteroids collide, in war when military armour is penetrated, when you spray a mist of water over your beloved plants, and yes, when an inflated balloon is popped.

In a popped balloon, the number of pieces you end up with depends on the stress inside the rubber wall. This stress comes from the difference between the air pressure inside the balloon and the air pressure outside it. The higher the difference in air pressure, the higher the stress in the rubber. If you then stab a hole in the inflated balloon, the air escapes and the stress in the stretched rubber is suddenly relieved.

Suppose that the balloon is only moderately inflated. The stress in the rubber is low. When the balloon is pricked, a single crack will start at the hole, and then race around the balloon, usually splitting it into just two fragments. This single crack is enough to relieve the stress.

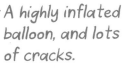

Blade

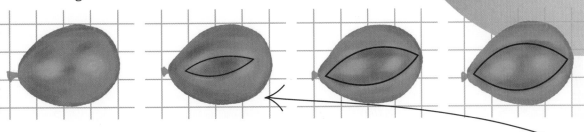

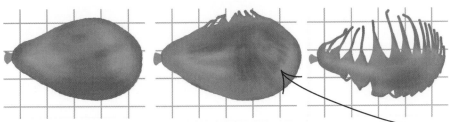

But if the balloon is stretched very thin with lots of air, and then pricked with a pin, the result is very different. Starting from the puncture point, a bunch of cracks expand out radially, like the spokes on a bicycle wheel. These cracks split into more cracks as they spread outward. The French scientists who did this research got a maximum number of 67 pieces from a fully inflated balloon. In general, all the cracks seem to be separated by roughly the same distance, and they end up looking a little like branches on a tree, or a rib cage.

A highly inflated balloon, and lots of cracks.

A moderately inflated balloon and just a single crack.

The higher the stress in the rubber means that there's more stress needing to be released. But the rubber can't tear any faster. The only way for the rubber to relieve the stress is for the crack to split into more cracks, which then split into even more cracks. By vastly increasing the number of cracks, the stress on the rubber is released more quickly, and the air inside gushes out.

Who said cracking up under stress was something that happens only to humans? Balloons do it, too!

ASTEROID DODGING

Personally, I love a bit of 'high drama'. So here's a friendly reminder that we are mere mortals who could be wiped off the planet in a jiffy.

Though, when I say that 'humanity had a lucky break on Halloween, 31 October 2015', I'm not exaggerating – not even a little bit. And most people didn't even know we nearly got wiped out by an asteroid.

ASTEROIDS 101

Out there in our Solar System are small, rocky objects orbiting the Sun, called asteroids. They're leftovers from when the planets formed.

One of these asteroids – a big one, about 10 km across – collided with the Earth around 65 million years ago, leading to the extinction of the dinosaurs. (Most of the dinosaurs, anyway; I know that some of them survived. Today, we call them birds.)

Much more recently, on Halloween in 2015, an asteroid (2015 TB145) just missed hitting our planet. It was about 650 m across. Yes, it was a lot smaller than the dinosaur buster.

But regardless of the size difference, a rock hitting the Earth at some 30 km/sec would cause a lot of damage to the human race.

How much damage any fast-moving rock 650 m wide would cause depends on where it lands. If it landed on a potential super-volcano site such as Yellowstone National Park in the USA or Mount Etna in Italy, it could set off volcanic eruptions blackening the skies worldwide. Or if it landed in the mid-Pacific Ocean, it could set off massive tsunamis. Tsunamis and volcanoes are bad enough, but if the asteroid landed on a city, then most of the people in that city would die.

We know these rocks are out there, so how do we protect ourselves?

Well, we could gently nudge asteroids away from hitting us. The radius of the Earth is about 6,400 km, so that's the maximum distance we would have to shift an incoming asteroid off course. If we had three years' advance warning, we humans could band together, develop the necessary technology, and do it.

DETOUR

18412
Kruszelnicki

I have my own asteroid. It's called 18412 Kruszelnicki. It's about 6 km across, and is in an orbit between Mars and Jupiter.

The famous astronomer Robert McNaught discovered it way back in 1993. Because he discovered it, he had naming rights, and he very kindly named it after me.

In 2006, he discovered the brightest comet visible from Earth in several decades. The Great Comet C/2006 P1 is also called Comet McNaught, and the Great Comet of 2007. Our family went to The Gap at Watsons Bay in January 2007 to look at this comet as often as we could. Looking westward over the city of Sydney, we could clearly see Comet McNaught at sunset. It was magnificent.

61

31 Oct 2015

31 Oct 2015

But don't wipe the sweat off your brows just yet. It missed us, but that doesn't mean that we have nothing to worry about. There are many more civilisation-busting asteroids out there. And we don't even know where they are. We think that fewer than half of the estimated 2,400 near-Earth asteroids between 300 and 500 m across have been discovered. Why?

Part of the reason is that we spend more money making movies about asteroids hitting the Earth than we do looking for these asteroids. I'm a huge fan of space movies, but I reckon we should also do the Science as well.

The other thing is that they can be hard to find. For example, this asteroid, 2015 TB$_{145}$, spends most of its three-year orbit moving quite slowly out past Mars – and only a small amount of time in the inner Solar System, looping around the Sun at high speed. In addition, its orbit is highly tilted relative to the plane of the planets (about 40°).

This was 2015 TB$_{145}$'s first recorded visit. So our first proper look at it happened as it zipped past at a very scary and uncomfortably close distance. But because it was so close, we got some very good pictures. Using some very big radio telescopes, we found that it rotated every five hours.

Given that it zipped past on 31 October, most of the media coverage of 2015 TB$_{145}$ was wrapped up and sugar-coated with a Halloween angle. The asteroid turned out to be roughly spherical, with a few bumps and craters, so at a stretch you could compare it to a skull. Trying to be cute, the media nicknamed TB$_{145}$ 'Great Pumpkin' or 'Skull Asteroid'.

But if 2015 TB$_{145}$ had actually hit Earth, we would have been struggling with real corpses, not playing ghostly dress-ups. 2015 TB$_{145}$ is the Universe's way of poking us and asking, 'Hey, humans, how's your Space Program going?'

I really think that we humans have to become a space-going race. It's not just to avoid destruction from low-flying asteroids. There are all the other benefits that come with new technology, most of which can't even be imagined yet.

Earth has been our Cradle of Life, but as we grow, we might have to leave home. Especially before we get stuck between a rock and a hard place …

DEFENDING EARTH

In 2022, NASA will try to redirect the path of a flying rock, which is roughly the size of the Great Pyramid of Giza. We're taking baby steps. This will be our first practice run – and on an asteroid that will definitely miss us.

Around October 2022, an asteroid called 65803 Didymos will pass about 11 million km from the Earth. Like about 15% of all asteroids, Didymos is actually a twin asteroid system. The primary asteroid (Didymos A) is about 780 m in diameter. It has a small satellite (Didymos B), charmingly nicknamed Didymoon, which is about 160 m across, and which orbits about 1.1 km away from Didymos.

The test is called DART – Double Asteroid Redirection Test. DART carries relatively few science instruments. Its main job is to smash into Didymoon with its 500-kg mass, at a closing speed of about 6 km/s (about 22,000 kph).

It is estimated that this impact will change the velocity of Didymoon by about 0.4 mm/s. This is a very tiny amount. The exact position of Didymoon will also change by a very tiny amount. But as the asteroid travels along its new orbit, over several years, the cumulative changes will add up.

Originally, this was to be a joint collaboration between NASA and the European Space Agency (ESA). Because the ESA contribution was cancelled, there won't be a nearby spacecraft to monitor the impact. So the immediate effects of the impact will have to be monitored from Earth, with ground-based telescopes and radar.

This will be effectively humanity's first planetary defence test against asteroids.

COMING CLEAN
ON CLOTHES

Nowadays, the hardest part of washing your clothes is deciding if they're dirty enough to need it. If the answer is 'yes', you dump them into the washing machine, add some detergent, press the magic start button, come back an hour or so later – and they're done.

You're happy because the clothes are wet and smell clean, but scientists want more – they want to explain what forced the dirt out of the 'pores' in the clothing.

Believe it or not, this is the not-so-famous 'stagnant core' problem. It's been a problem for several decades in the field of 'laundry detergency'. (Do you like the word 'detergency'? Honestly, it's real.)

But in 2018, after thousands of years of washing clothes, we finally found the scientific answer as to how washing works. It's actually not as simple as you would have guessed. Just agitating the clothing in water is not enough to wash out the dirt.

Science now tells us that the freshwater rinse creates chemical and electrical imbalances. These imbalances then drag the dirt out of tiny crevices in the threads.

Problem solved and washing cleaned. Win–win!

CLOTHES 101

Clothes, of some kind, have been around for a long time.

When our small-brained ancestors split off from the chimpanzees about 7 million years ago, they definitely had a thick protective coat of fur. Some time around 2 to 3 million years ago, we began to lose our fur, thanks to a mutation. It gave us a bunch of advantages and disadvantages.

One benefit to us was that lice and fleas now had a much smaller potential hairy home on our bodies.

We could also get rid of body heat much better. So thanks to having less insulating body fur (and more sweat glands), our

Textiles and urine

The first clues of manufactured textiles date back some 7,000 years, to ancient Egypt and what is now Switzerland. I'm impressed.

In early civilisations, individuals and small groups would wash their clothes in the nearest creek or river. The dirty clothes would be beaten over rocks, pounded with feet or wooden bats, or scrubbed with stones.

The textile industry was in full swing in Roman times, about 2,000 years ago. The Romans had an entire industry devoted to cleaning clothes – the 'fullers'. The fullers' shops were physically large, and usually one of the main employers within the city.

The fullers would dye, wash and dry all types of clothing. They would use a soap made from fat and ashes, as well as human urine, to wash the garments.

Back then, human urine was a commodity that would be not only collected from public toilets within the city but also sometimes imported. It was so valuable that the government taxed it.

ancestors could run very long distances to chase down their next meal without getting overheated.

But there's another surprising advantage of not making lots of fur. Fur is made from protein. If you're not putting your protein into fur, you can put it into a bigger brain. Perhaps that mutation of losing our body fur made us into the clever folk that we are today.

But one disadvantage of not having fur is an increased risk of skin cancers. Certainly, we began to evolve darker skin around 1.2 million years ago, as a protection against skin cancers.

Another problem is that to help us keep warm on cold nights, we now needed clothing – or lots of furry friends to lie next to. Back then, we didn't have a textile industry, so our first clothes were fur from animals.

But clothes get dirty. After the Dark Ages in Europe, washhouses, where groups of people would gather to wash their clothes, became common. And laundromats are still around for communal clothes washing today.

WASHING MACHINES

Washing your clothes can definitely make them cleaner. But washing clothes by hand was hard labour. Washing machines were a very attractive alternative.

One design from the 1670s for a washing machine involved putting the laundry into a bag. It would then be soaked in water, before being squeezed by a 'wheel and cylinder' mechanism. One of the earliest English patents for a washing machine was issued in 1691.

Less than a century later, in 1787, Edward Beetham ran an advertisement in *The Times* on 10 October. The ad claimed that he had 'a machine for washing linen which will, in an equal space of time, wash as much linen as six or eight of the ablest washerwomen, without the use of lees [lye, a major component of soap], and with only one third of the fire and soap'.

By the mid-1850s, commercial laundry machines driven by steam were available in the UK and USA.

Washing machines powered by electricity were first advertised in the early 1900s. By 1940, 25 million homes in the USA were wired for electricity – and 60% of them had an electric washing machine.

But washing machines did not become common in Europe and the UK until the 1950s, because of the economic devastation caused by World War II.

REST AREA

CLEANLINESS IN FRANCE

Every society has its own standards of cleanliness and hygiene – and they change over time.

Professor Georges Vigarello discusses this in his book *Concepts of Cleanliness: Changing attitudes in France since the Middle Ages.*

Vigarello points out that in the city of Paris alone in 1292, there were 26 public baths. They were places of leisure and relaxation, like taverns.

But by the 1400s, bathhouses and steamhouses began to fall out of favour – mainly for moral reasons. In 1566, a law in the city of Orléans ordered the shutting of all places of prostitution, including bathhouses. Within a few decades all public steamhouses were shut in France. By 1692, the handful of public bathhouses left were reserved for only medical uses.

Thanks to the recent invention of the microscope, water was at that time seen as 'dirty' and full of unwelcome creatures. Water was thought to enter the pores and make the internal organs 'fragile'. Water was no longer used for personal hygiene, apart from rinsing the mouth, hands and face. This led to the concept of 'visible purity', which meant keeping only the hands and face clean – because they were on show. But kind of illogically, people still washed their clothes.

Around the mid-1700s, bathing started to become more common. But bathing was not seen as hygienic, but more as a luxury for the elite, or as a tonic to strengthen the internal organs if taken cold.

Only in the mid-1800s, with the rise of the concept of 'filth' as a cause of disease, did bathing get promoted as an act of positive hygiene.

Environmental impact of clothing

Everything we do in some way disturbs the Universe. So, as a part of the Universe, washing clothes does have an impact.

The actual manufacturing of the garment, and then its packaging, transportation, use and disposal, accounts for less than half of the environmental impact.

The bigger environmental impact comes from the act of washing the clothes. To take one example, about 60% of the environmental impacts of Levi's jeans comes from the repeated washing. This includes water use, detergent, heating, and disposing of the grey water.

Intrayarn permeation – tiny, tiny gaps or crevices, about 1 μ in diameter (that's 1 micron, which is 1 millionth of a metre), where dirt hides and water can't easily get into

Interyarn permeation – big pores or holes (200 u) that water easily gets into and out of

ANATOMY OF GARMENTS

Over the last century, science and engineering have combined forces to make washing machines better.

However, a small group of scientists were troubled (thanks to a worrying 'glitch' in the Land of Washing Theory). They could never explain how dirt was being removed from clothing.

The problem was that using their standard scientific knowledge, the dirt would take more than six hours to drift out of the clothing, to then vanish away with the rinse water. That's obviously a problem, because your clothes get clean in an hour-or-so. The problem had even been given its own special name, the Stagnant Core Problem.

The scientists knew that dirt got trapped inside a little crevice (or a hole, or a core) inside the clothing. Chemistry told them the detergent molecules grabbed onto the dirt. But they could not explain what force pushed the combination of dirt and detergent out of its cosy little crevice or core. The Laws of Diffusion predicted that the dirt and detergent should simply remain, or stagnate, inside this little crevice or core – hence the name, the Stagnant Core Problem. But that couldn't be correct, because clothes would never come clean. So what were they missing?

To find the answer, let's take a close look at fabric.

Clothing is made of material that has many threads or yarns. Often, half of these threads are parallel to each other, and they often cross the other half of the threads at right angles. They can loop up and down as they cross the other threads or yarns. So if you look at one particular thread, you can see that it might go over one crossing thread, then under the next crossing thread, then over and under and so on until you get to the very edge of the fabric.

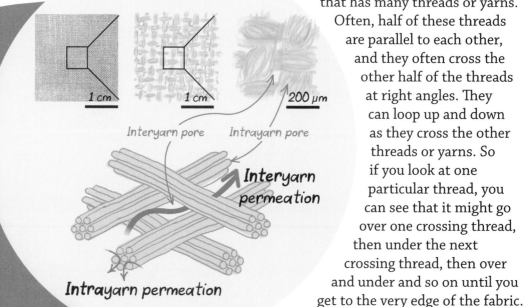

Interyarn pore   Intrayarn pore

Interyarn permeation

Intrayarn permeation

This creates a lot of friction between the threads, which has the benefit that they don't unravel.

The regular pattern also creates many tiny holes, each roughly square, and bounded on four sides by four threads. These holes are about 200 microns across – about three or four times the thickness of a human hair.

These holes are called the interyarn pores (see the diagram opposite). They are very permeable. Water can easily rush through these relatively large holes while your clothes are being cleaned in your washing machine.

THE STAGNANT CORE

But now we're nearly at the stagnant core – which is the problem.

Let's take a closer look at the actual thread or yarn. It's made up of (say) half a dozen smaller fibres that are wound very closely together. As a result of the tight winding, there are tiny gaps (or crevices) between these very small fibres. These crevices are open at the top, but closed on the inside. These are the intrayarn pores (see diagram). So they have a blind end. They're deep and skinny – as narrow as 1 micron across.

And this is where the dirt hides. This is the 'core' that is 'stagnant'. There's virtually no liquid movement inside this core.

The dirt can get in because the thread is just slightly moist, and the dirt slides in with a bit of help from capillary action. If some washing water does get into this crevice or core between the fibres, it can't push the dirt out through the other end, because it's closed off.

Next, let's add some detergent. The detergent molecule is very special. One end loves fat, so it sticks to the dirt (which is usually fatty). The other end loves water, so it hangs out waving gently to-and-fro in the water.

Now using the Laws of Diffusion, it would take at least six hours for most of the dirt to slowly diffuse from these stagnant cores.

But in reality, it takes much less time to do a load of washing. So, what gives?

DOWN A GRADIENT

It wasn't until 2018 that Dr Sangwoo Shin, an Assistant Professor in Mechanical Engineering at the University of Hawaii, solved the issue.

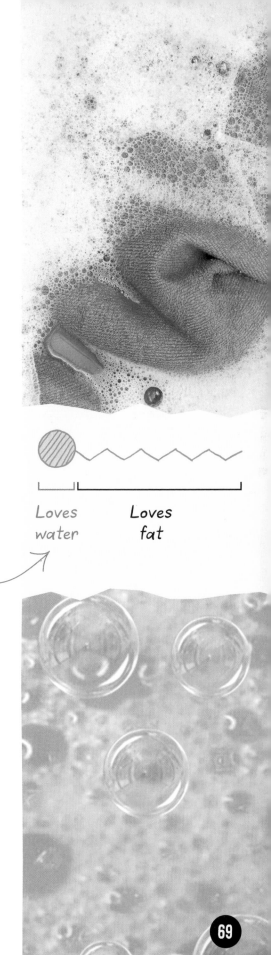

Loves water

Loves fat

Complex fabric

Fabric technology is quite complex. Not all the fibres are parallel, and they don't all cross at right angles. And they can have special coatings, for special applications, that affect the look, feel and function of the fabric.

Dr Shin had an advantage in trying to solve how stuff gets out of these cores. He had previously written about 'diffusiophoresis' in 2016. This is a made-up word, from 'diffusion' and 'electrophoresis'.

If you study diffusiophoresis, you're looking at one type of diffusion caused by a difference in electric fields (electrophoresis), along with another type of diffusion caused by differences in concentration gradients (chemophoresis).

His previous work had used the concept of diffusiophoresis to insert chemicals into these blind holes that had only one opening. He had worked on a kind of gradient, to push chemicals into dead-end channels. This work had very many applications, all the way from drug delivery in the human body, to recovering underground oil and gas.

The basic concept behind a gradient is that 'things' or 'processes' will normally go down a gradient.

A ball will roll down the gravity gradient of the hill, from top to bottom.

Dye will diffuse down a concentration gradient. Try putting a tiny droplet of red food dye into a bowl of water – and watch what happens. The concentration of red food dye molecules is very high inside the actual droplet. At the edge of the bowl, the concentration of red food dye molecules is 0%.

But if you wait a while, the molecules that make up the red food dye will go down this concentration gradient – and spread out to the edges of the bowl. Eventually, you end up with the whole bowl of water having the same concentration of food dye molecules – probably a very pale pink.

Water with detergent

Very slow removal of balls (or dirt)

Minutes:
0   1   3   5   10

Slow removal of balls (or dirt) from top of column only

EXPERIMENT TIME

So Dr Shin carried out a few experiments.

First, he made up a whole set of blind-ended narrow channels (about 48 microns across). And then he put even smaller particles inside the channels (fluorescent polystyrene balls about 0.5 microns across). He added detergent to the water and some of it got into the narrow channels.

And very, very slowly, the tiny particles began to drift out. Just leaving the clothes in the detergent-soaked water took some dirt out – but very slowly.

His next step was the equivalent of a clean rinse cycle. He took away the detergent water above the narrow channels, and replaced it with fresh water. Suddenly there was a concentration gradient of detergent.

There was lots of detergent inside the narrow channels, and none in the fresh water above them. So the detergent started moving down the concentration gradient. It started leaving the narrow channels. And because the detergent was stuck onto the tiny dirt particles, it dragged them along as well.

He did the same experiment in a few different ways, but the result was the same. Fresh water set up a gradient that pulled the particles out of the narrow, blind-ended channel.

He had proven, after adding detergent, that 'rinsing with freshwater is the key to effective cleaning'. It's the sudden surge of the fresh water rinse that drags the particles of dirt out of your clothing. But this fresh water does not get into the crevices.

No.

The fresh water stays outside the crevice, and drags the dirt up (out of the gutter!).

In fancy sciencey talk, the process of 'diffusiophoresis' kicks in, and a combination of electrical gradients and chemical gradients expels the dirt from the crevice.

This new understanding could lead to better washing machine cycles. Dr Shin said, 'Since the current finding suggests that rinsing is the key player in dirt removal, we can design a better wash-rinse cycle for optimal cleaning'. Perhaps we have fewer rinse cycles, but they could be longer. These insights could also lead to new detergents, and to improvements in other industries where you want to shift small stuff.

Now that we know how the dirt goes, all we have left to solve is, where do the odd socks go?

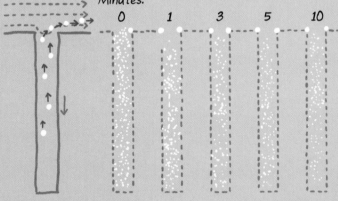

Plain fresh water

Much faster removal of balls (or dirt) from throughout length of column

Minutes: 0   1   3   5   10

Quick removal of balls (or dirt) from most of column

WHY ELEPHANTS DON'T GET MORE CANCERS

A 'paradox' is a statement that seems to be absurd or self-contradictory but may still actually enlighten us.

Here's a paradox that I hope will shine a light – Peto's Paradox. It was proposed way back in the last century, in 1975, by, you guessed it, Richard Peto. Back then, he was a young statistician at the University of Oxford.

Peto's Paradox ponders why humans suffer from so few cancers – at least, compared with mice, which are much smaller than us and have vastly fewer cells than we do. If the probability of getting cancer was simply related to how many cells an animal has, one would expect humans to have a higher incidence of cancer than mice, right? Well, quite the opposite is true, and herein lies the paradox. And this paradox gives us an unexpected clue as to how to reduce cancers in humans!

PETO'S PARADOX

But hang on – we humans get lots of cancers. One in every three of us will suffer some kind of a cancer at some stage in our lives, and, unfortunately, about one in every five of us will die from a cancer.

But compare humans with mice. We are a lot bigger than mice, and so we have many more cells. That means more cells will divide into their daughter cells. In addition, we live a lot longer than mice. If you factor in both the extra cells and the extra years, our human cells do about 30,000 times more divisions than do all the mice cells.

But cell division is a risky process. Whenever a cell divides, there's the chance of a mistake being incorporated into the new DNA – and, sometimes, that mistake could increase the chances of cancer. So even though we have many many more cell divisions than mice, both humans and mice have roughly the same lifetime risk of getting a cancer. But statistically, we humans should have 30,000 times more cancers (based on cell divisions).

So Peto's Paradox is simple to state: we humans have many more cells, and we live longer, than mice. So we *should* have more cancers. Why don't we?

And if you scale up to elephants, which are both long-lived and huge, they *should* be dying from colon cancer at the tender age

What is cancer?

The first thing to realise is that 'cancer' is not an external 'thing' that invades your cells. Yes, sometimes it might be caused by an external factor, such as smoking tobacco or breathing in petrol vapour when you refuel your car. But sometimes a cancer happens for an internal reason. Your own cells might stop obeying the commands that tell them to grow to a certain size and number – and then stop. Instead, they just keep on growing.

So, a cancer is simply your own cells doing the bad thing of keeping on growing when they should have stopped at a certain stage.

Second, 'cancer' is not a single disease. There are hundreds of different cancers in humans. For example, in the lungs, there are four main cancers. Three of these are caused by smoking, but one is not (we don't know what causes it).

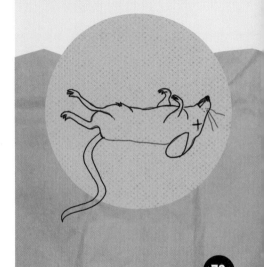

Age and cancer

One of the strongest influencers of cancer rates in humans is age. The chance that a person will get a cancer in the next five years of their life is only one in 700 if the person is 25 years old. But it's one in 14 if they are 65 years old. That's an increase in risk of 50 times!

Cancers happen more often as we get older. One possible reason (there are so many!) is because a longer life means a longer exposure to cancer triggers (or carcinogens). These triggers can be both external background, and internal spontaneous, factors.

of three. But exactly the opposite is true! Something is somehow protecting elephants from cancers. Only one in 20 elephants die of a cancer – even though they weigh 100 times more than we do. And their bigger elephant body is made of lots of extra cells, which undergo lots of extra cell divisions. What is going on?

Why do only 5% of elephants die of cancer, while the figure for humans is 20%? Why do we get more cancers than elephants?

'SUPERHERO MOLECULE OF THE YEAR [1993], p53'

So maybe elephants (and perhaps other large animals that live for a long time) have special biological 'weapons' that fight cancer. If that's the case, what are they?

Nearly half a century after Richard Peto first proposed it, we might be close to solving Peto's Paradox. The answer seems to be related to a protective gene called p53 – which, among many other activities, appears to help fight cancer.

It's a very old gene found in all multi-cellular animals. It finds damage inside a cell – and then stops that specific cell from dividing until the damage is repaired; or if the damage is too great, it forces the cell to self-destruct.

We humans inherit one copy of the gene p53 from each parent – two copies in all. But elephants have 20 copies. When elephant cells are exposed to cancer-causing radiation, their cells are twice as likely as human cells to self-destruct. If the cell dies, the cancer is stopped dead in its tracks. (Almost certainly, the extra copies of the gene p53 are the reason.)

Some people are unlucky enough to have only one 'good' copy of the p53 gene. They suffer from a rare condition called Li-Fraumeni Syndrome. They have one defective copy of the p53 gene, as well as one functioning copy. They usually get cancer while in childhood, sometimes when they're as young as two years old. Their lifetime risk of getting a cancer is close to 100%.

So what does this discreetly named gene actually do? The p53 gene makes a protein, also called p53 (that bit seems straightforward). The protein was discovered in 1979

CELLS AND RISK

About 600 million years ago, creatures evolved from single-celled creatures into multi-celled creatures.

But here's something strange. Overall, the risk of cancer in the cells of a living creature does not increase with the number of cells in that creature's body.

And then it gets more tricky again.

Sure, we humans carry about 37 trillion cells that arose from our original single fertilised egg (from when our parents loved each other very much, in a special way). But each day, the number of cells that divide to make new cells in the human gut is about 100 billion. That is a huge amount of turnover – new cells being born, old cells dying, the overall number staying the same.

This turnover in the gut is also about 300 times less than the number of cells in the body. But after 300 days (roughly a year), our gut alone has made as many new cells as there are cells in the whole human body.

Maybe this huge turnover – with the constant risk of mistakes being made – is why anti-cancer mechanisms have had to evolve in so many different animals.

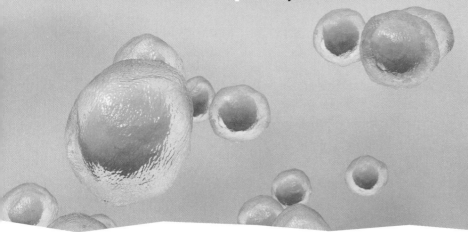

Warning – confusing names

p53 is the name of a *gene* in your DNA. It's made from various nucleotides – the well-known Adenine, Thymine, Cytosine and Guanine.

I am sorry to tell you that p53 is also the name of the *protein* made by the gene called p53. This protein is made from various amino acids – which are a very different chemical from nucleotides.

However, there is some way out of this confusion.

Since being discovered, the protein has been given many names besides plain old p53. These include Cellular Tumour Antigen p53, Antigen NY-CO-13, Phosphoprotein p53, Tumour Suppressor p53 and Tumour Protein p53.

So here's the solution. First, we pick an extra name and call it TP53.

Second, when we write about the 'gene' in a science journal, we write the name in italics, e.g. *TP53*. And when we write about the protein in a science journal, we use regular typeface, e.g. TP53. But in this book, we will just use p53, and I'll let you know whether I'm referring to a gene or a protein.

Subtle, huh? And how do you *speak* in italics? 😉

by oncologist David Lane and colleagues. It got its name because it is a protein (hence the 'p'), and it has a molecular weight of 53,000 daltons (ditto '53'). Just for comparison, that's about ten times heavier than an insulin molecule.

It's a very famous molecule, this p53 protein. It's had over 15,000 papers published about it, and in 1993, was declared Molecule of the Year by *Science* magazine. It's often called the Guardian of the Genome. It protects our DNA.

Inside the human body, p53 has all-powerful properties – it decides if a cell is to live or die. (Maybe it should have been subtitled 'The Great Decider'?)

P53 - THE GREAT DECIDER

When damage to the DNA is detected, one p53 protein can spring into action to join up with three other proteins of p53. Then the Group of Four will try various actions.

First, it will try to repair the damaged DNA. Sometimes it can, but sometimes it cannot. When it can't fix the DNA, the p53 Group of Four can force the cell to commit suicide. This ability to force damaged cells to die is one way that p53 protects us from cancer.

But the p53 gene is not perfect. This powerful guardian can sometimes make mistakes by itself (and, for example, start attacking our own bodies). And furthermore, we humans can damage the p53 gene with cigarette smoke and ultraviolet light.

CANCERS AND P53

Today, we know that about half of all human cancers show some kind of mutation in the p53 gene. This has led to some helpful therapies.

A few preliminary studies have shown that repairing the p53 gene eliminates some cancers.

Another therapy involves leaving the faulty p53 gene in the DNA alone. Instead, it concentrates on repairing the faulty p53 protein – and this therapy has had some success.

Another (and more subtle option) is to bypass the p53 gene and the p53 protein completely. Instead, go one level down. Go to the processes that the p53 protein normally activates, and switch them on artificially.

YES AND NO

There is a down side to having p53 protecting us.

Although the p53 molecule can limit cancer, unfortunately, it can also accelerate ageing.

I know, this sounds like another paradox – after all, aren't ageing and cancer linked? How is it possible that the opposite happens? How can ageing increase, while cancer is decreasing?

The answer lies in stem cells.

One pathway of ageing is that we lose the ability to repair and regenerate damaged cells. More specifically, we run out of the stem

AVOIDING CANCER

Different animals have evolved different strategies for avoiding cancers.

The bowhead whale can weigh up to 100 tonnes, and can live longer than 200 years. Its anti-cancer strategies involve changes (mutations) in the genes related to ageing, and to DNA repair. These whales have both mutations and extra copies of these genes.

The much smaller naked mole rat lives for an unusually long time for such a small animal, and also seems to avoid cancer. Its cancer avoidance technique has at least two aspects: one limits cells sticking together (when a cancer starts growing, lots of cells start sticking together); while the other involves cell division (when a cancer starts growing, the cells relentlessly keep on dividing, without stopping). The mole rat possibly has another anti-cancer technique that involves 'inflammatory responses'.

There are almost certainly many other anti-cancer strategies used throughout the animal kingdom.

cells that can do this repair and regeneration. And why do we run out of stem cells? Because these stem cells become damaged, and are then forced to commit suicide by the p53 response pathway.

If the p53 system activates too readily, we get less cancer – but we age more rapidly. If the p53 system activates too slowly, we get more cancer – but we age more slowly (because we have more stem cells left available to do repairs).

You can see there are incredible subtleties around p53.

But if we can understand them, like elephants, we might be able to both live longer and avoid cancers. Maybe we should switch p53 on to maximum sensitivity every now and then just for a few weeks to clear out the cancer cells – and then drop its sensitivity down for the next few years?

But what if that forces us down another pathway? In this different non-Peto Paradox, should we choose to live longer only to get older quicker?

CANNIBALISM – A NUTRITIOUS DINING OPTION?

Nutrition Facts

Serving Size ...g
Servings Per Container

Amount Per Serving

Calories ...   Calories from Fat ...

% Daily Value*

Total Fat ...g   ...%
  Saturated Fat ...g   ...%
  Trans Fat ...g
Cholestrol ...mg   ...%
Sodium ...mg   ...%
Total Carbohydrate ...g   ...%
  Dietary Fiber ...g   ...%
  Sugars ...g
Protein ...g

Vitamin A   ...%
Vitamin C   ...%
Calcium   ...%
Iron   ...%

*Percent Daily Values are based on a 2,000 calorie diet.
Your Daily Values may be higher or lower depending on your
calorie needs.

Lots of animals eat others of their own species. We humans are no different when it comes to cannibalism.

It seems we've been eating other people for nearly a million years. And yet (and this might be a good thing) people are not especially nutritious, especially when compared with animals of a similar size. So why eat your fellow human?

HISTORY OF CANNIBALISM

Life is full of little surprises. For me, it was not such a pleasant surprise to find out that, until the late 18th century, eating various bits of the human body was widely accepted as an approved medical therapy.

These 'bits' included flesh, bone, blood and even various types of moss that would invade human skulls after burial. Flesh from a mummified body was thought to be an excellent remedy for bruising. Indeed, powdered mummy, known as 'mummia', was still included in the therapeutic Merck Index (a medical index of chemicals, drugs and biologicals) until the early 1900s.

The Danes thought that drops of warm blood from a hanged criminal would cure epilepsy. This belief traces back 2,000 years to ancient Rome, where the blood of wounded gladiators was thought to cure epilepsy.

We've found pretty good evidence in caves in Spain of our distant predecessors practising cannibalism 800,000 years ago. And in popular culture, you might remember the fairytale children of *Hansel and Gretel*, with the evil witch fattening Hansel up so she could eat him. (That's assuming you think that witches are human.)

Cannibalism has happened in most parts of the world – and not just in so-called primitive or tribal societies. When people are starving, they'll eat anyone. In the USA, the Donner Party was a westward expedition in 1846 and 1847. By bad luck, they found themselves stranded by snow

What you're used to

US author Jared Diamond describes how some of his Indigenous friends in New Guinea were quite comfortable with their ritual cannibalism. Yet they were horrified at his descriptions of circumcision, and America's funeral customs and especially its treatment of the elderly, who are sent to nursing homes.

DETOUR

Cannibalism and disease

Back in the 1950s, a rare neurodegenerative disease called 'kuru' was annually killing up to 2% of the population of a small remote highland tribe in New Guinea, the Fore. The tribe practised ritual cannibalism – as an act of grief, or love. Kuru was spread by eating the brains of somebody already suffering the disease. Kuru was an infectious disease, but virtually impossible to catch if you didn't practise cannibalism. It mostly affected women and children, but not many adult males – who did not join in much at the feasts.

Kuru is caused by an infective agent called a 'prion'. It's a protein that has become misshapen. It's quite different from a bacterium, virus or parasite. Kuru is similar to Creutzfeldt-Jakob Disease, which occurs worldwide at the rate of one per million people per year. Other prion diseases include scrapie in sheep, and Bovine Spongiform Encephalopathy ('Mad Cow Disease') in cows.

in a high mountain pass in California – and the survivors lived by eating human flesh. There were many cases of cannibalism in World War II, most notably in prisoner-of-war camps, and also during the Siege of Leningrad.

In 1972, Uruguayan Air Force Flight 571 crashed in the Andes. The survivors were rescued after 72 days, but they had already desperately resorted to eating those who had already died.

And some people practise 'placentophagy'. This is the eating of the placenta, or afterbirth, once a child has been born. This could count as a type of cannibalism.

RITUAL CANNIBALISM

One classification of cannibalism breaks it down into 'endocannibalism' (where a person from within the community is eaten) or 'exocannibalism' (the consumption of a person from outside the community).

Endocannibalism is usually associated with rituals. These include the grieving process after the death of a tribe member, or a pathway to make sure the souls of the dead make it into the bodies of the next generation. Endocannibalism was a mark of respect and honour – either to the dead that were eaten, or the living who were doing the eating.

On the other hand, exocannibalism was usually associated with hostility, violence and contempt. It was often a celebration of victory over the enemy.

In both endocannibalism and exocannibalism, there is some degree of belief that the eating of another's flesh will give the cannibal some of the victim's desirable characteristics.

MOTIVATION OR USE OF CANNIBALISM

Another way to describe cannibalism is to consider its function or motivation.

'Ritual' or 'magical' cannibalism relates to religious beliefs. 'Survival' cannibalism happens as a last resort to keep at least some of the group alive. And of course, there is 'Psychotic' or 'Criminal' cannibalism (think Hannibal Lecter). 'Nutritional' cannibalism looks at eating flesh in terms of its nutritional value or taste.

NUTRITIONAL VALUE OF HUMANS

Dr James Cole is a Principal Lecturer in Archaeology. He is especially interested in human evolution. In 2017, he wrote a paper on the topic of how many calories/kilojoules you could get from eating people. To get calorific values for the human body, he used papers written in the 1940s and 1950s. These papers used a very small sample size of just four human adult male cadavers. Even so, it's probably not a bad first approximation (or guess).

First, while most animals are about 60% muscle, and fish and birds are about 80% muscle, we humans are only around 40% muscle.

So, to keep it simple, let's compare how long you would be kept alive by eating humans, or other animals.

Let's start the comparison with mammoths, which were hunted to extinction about 5,000 years ago, but which were around for most of the 300,000-year history of *Homo sapiens*. The skeletal muscle of one mammoth offered about 15 million kJ of energy. Then compare the mammoth to a horse, which delivers about 840,000 kJ, while a human has a potential of only 134,000 kJ.

A mammoth would keep a modern human nourished for over 1,700 days; a horse would keep you going for 96 days; but eating another human would give you only 15 days. In addition, it was probably safer to chase down an animal than another intelligent human – you might not survive an encounter with another human.

So why bother eating another human if this was a potentially risky strategy, and, relatively speaking, there wasn't a huge amount of nutrition available at the end of it?

Perhaps nutrition wasn't the only reason we humans have such a long history of cannibalism. We can only speculate as to why this was so. After all, all types of humans (including Neanderthals) had quite complex attitudes to the burial of their dead. Perhaps the social significance of cannibalism was another factor?

Over time, in Western society, cannibalism has become taboo. Almost, you could say, in poor taste …

Anthropophagy

The root 'anthrop' means 'people', while 'phagy' means 'eat'. So 'anthropophagy' is the fancy word for 'human cannibalism'.

Two cannibals are eating a clown. One says to the other: 'Does this taste funny to you?'

BARCODE INVENTION

Every single day, some 5 billion barcodes are scanned around the world. That's almost one scan for each person on Earth!

And it's not just your groceries. It's everything from airport luggage and wristbands for newborn babies, to livestock tags, library books, bags of blood for transfusions, registered mail, movie tickets, pharmaceutical drugs and so much more. They each have a barcode, begging to be scanned.

Translating the barcode from Thought to Action took a quarter of a century, and many twists and turns. The New Jersey Mob, elevator music and a university drop-out on a beach running his fingers through the sand each played a part.

The evolution of the barcode itself didn't follow a simple path. In fact, the barcode never started as the familiar rectangular set of parallel lines so ubiquitous today. Instead, it began as a circular 'bullseye', with sets of nested circles inside one another.

KNOWLEDGE IS SALES

In the early 1980s, the then CEO of Pepsi-Cola, John Scully, asked Scott Klein (a Pepsi brand manager) to start analysing the store sales data. You see, there were now huge amounts of information that could be very easily gathered – thanks to the barcode on each can of Pepsi.

The early results were entirely expected. Sales of Pepsi increased when there was a sales promotion.

But closer analysis of the data showed a surprise – the sales increased the night *before* the sales promotion began!

It took a while to work out why. It turned out the stores would erect big displays of Pepsi-Cola drink inside the stores a day or two *before* the sales promotion began. It wasn't the lower price that was driving the sales – it was the fact that the product was being displayed in a big way. This knowledge changed how Pepsi marketed its soft drink in stores.

Another example is the American beef industry, which began barcoding beef quite late – after 2000. Before this, they thought that the biggest steak buyers were the wealthy. But after barcodes were introduced, the beef industry could link each slab of beef sold to a specific credit card. They were very surprised to find that rural blue-collar workers, with a fairly low median household income of around US$40,000, were their major customers.

Just like Pepsi, this specific knowledge led to a change in their marketing strategy.

MOB & MUSIC

The barcode story begins with our dreamer, N. Joseph Woodland. During World War II, he played a small part in the invention of the atomic bomb, working at the Oak Ridge National Laboratory in Tennessee. In 1947, he graduated from the Drexel Institute of Technology in Philadelphia with a degree in mechanical engineering.

Around this time, Joe – being a bit of an inventor – came up with a brilliant upgrade for playing music in elevators. (Yup, elevator music was big business back then.) The then-current system relied on just a few tracks of music coming from either records or reel-to-reel magnetic tape.

Joe's brilliant idea was to use 35-mm film, which could carry a whopping 15 tracks of music – five times more. He wanted to work full time on his brilliant idea. But his father in New Jersey warned him that the elevator music business was controlled by 'the Mob' (organised crime), and that he should keep away.

Joe obediently gave up his dream of lifting elevator music to the next level, and went back to Drexel to study a master's degree in engineering.

(Meanwhile, you have to wonder ... Would elevator music have evolved into something better if Joe Woodland had managed to pry the Mob's sticky fingers off it?)

CODING GROCERIES

In 1948, a manager of a small grocery chain, Food Fair, visited the dean of engineering at Drexel, begging for help. He needed an automated system that would stocktake supplies more accurately, but, most importantly, speed up sales at the checkout. You see, back then, either each grocery item was individually marked with its price, or the checkout person had to look up the price on a paper list. Groceries were typically lots of small-value items, and it took a lot of time to process each one. Grocery stores were missing out on sales, because they couldn't move the customers through quickly enough.

The dean said he wasn't interested.

But by an amazing coincidence, a fellow post-graduate student in engineering, Bernard Silver, overheard this conversation and excitedly reported it to his mate, Joe Woodland.

Joe was so convinced he could invent the solution that he dropped out of his master's degree at Drexler Engineering School. He went to Florida to live with his grandfather, so he could perfect his great invention. He survived financially by selling some stockmarket shares he had bought previously.

Joe hung around on South Beach, in Miami, thinking deep thoughts. What kind of code could capture all the information needed about a grocery item?

Well, the only code he knew of was Morse code, the dots and dashes he had learnt as a boy scout.

From the original patent

STROKE OF GENIUS

One day in January 1949, as Joe sat on a beach chair, inspiration struck.

'I remember I was thinking about dots and dashes when I poked my four fingers into the sand and, for whatever reason – I didn't know – I pulled my hand toward me and I had four lines. I said, "Golly! Now I have four lines and they could be wide lines and narrow lines, instead of dots and dashes." ... Then, only seconds later, I took my four fingers – they were still in the sand – and I swept them round into a circle.'

Joe had the idea to encode the information by varying the *widths* of the *spaces between* parallel lines or bars – wide or narrow. Morse code uses dots and dashes, digital code uses 1s and 0s, but this brand-new code proposed fat or thin bands. It would be easy to scale upwards – if you want to store more information, just add more lines.

By having these lines as concentric nested circles, the code could be read from any direction. This is because circles look the same, no matter which way you look at them.

And that's how the original barcode was born. It was a bullseye of circular lines, with the lines either narrow or wide.

ENGINEERS GOTTA BUILD

So now that they had their circular barcode, all that was missing was a way to read the darn things! For anybody who knows the Beatles' song 'Drive My Car', it was the same vibe. It was like having a driver, but no car.

So, Joe Woodland and his buddy from uni, Bernard Silver, built a system that could read their circular set of nested wide and narrow lines. Remember, this was the late 1940s, when electronic devices and primitive computers were expensive, big and clunky. So, using the best technology of the day, they eventually came up with a system about the size of a large desk that involved an ultra-bright 500-Watt light to shine through the paper and show the barcode, and an RCA935 photomultiplier tube to read it.

It kind of worked, but only just. It was unreliable and inconsistent. The basic problem was that they needed technologies that hadn't been invented yet. Lasers and cheap, powerful computers just didn't exist then.

Joe Woodland joined the IBM company in 1951, hoping to take the barcode and reader technology further – but that didn't really happen. Joe and Bernard kept their hopes up, and they were granted a patent in 1952 for their circular bullseye code.

It was eight years before the first laser was invented in 1960. The inventors of the laser got their Nobel Prize in 1964, but nobody imagined it would one day turn up in the supermarket checkout aisle. And meanwhile, following Moore's Law (that the number of transistors per chip doubles every two years), computers kept getting cheaper and more powerful every few years.

The grocers could do nothing but wait …

Culture and the barcode

Today, the barcode is so accepted that it has become embedded into our culture.

Barcodes are widely used by some protestors as an emblem of capitalism, or loss of identity.

Pink, the musician, has a barcode tattoo that includes numbers meaningful to her – her birthdate, the release date of one of her albums, her lucky number.

The cyberpunk TV series *Dark Angel* (2000–2002) was set in a dystopian distant future (2019!). Each person who had become a super-soldier via genetic modification had a barcode on the back of their neck.

The American artist Scott Blake has created artworks from barcodes, including 'Barcode Jesus: Flipbook', which was a digital mosaic made up of 1,536 barcodes.

BIG GROCERY ASKS, BUT ...

In 1966, one of the largest supermarket chains in the USA, the Kroger Company, begged for somebody (anybody!) to invent a way to scan grocery stock quickly and cheaply. Profit margins were dropping and labour costs were rising. They wrote in a little booklet, 'Just dreaming a little ... could an optical scanner read the price and total the sale ... Faster service, more productive service is needed desperately. We solicit your help.'

The Radio Corporation of America – RCA – set up a small team to respond to the challenge. They bought Woodland and Silver's patent, and tried to develop a scanning system based on the circular bullseye barcode. After six years of effort, they managed their first successful real-life store test on 3 July 1972, at the Kroger Kenwood Plaza store in Cincinnati.

There was a growing awareness in grocery stores across the USA of the need for some kind of Universal Product Code (UPC) for grocery items. This code could carry information, such as who made it and what the product was.

The shops needed a reader system that would recognise the UPC (whatever it ultimately turned out to be), and a computer system that would tie the product to a price, as well as special offers and discounts.

On the upside, by the early 1970s lasers could read a barcode and minicomputers were powerful enough to process the information stored in that code. The downside was that the barcode design was not yet finalised.

It was kind of a chicken-and-egg situation: the manufacturers wouldn't print barcodes on the labels if the shops didn't have scanners, and the shops wouldn't buy scanners if there were no barcodes to scan.

LET'S JUST DO IT

So a special committee was set up, the Symbol Selection Committee, to call for barcode/scanner proposals.

There were seven contenders.

RCA, having had real-life experience, were cocky and thought they'd win.

But, at the last minute, IBM put in a bid. They suddenly realised that Joe Woodland had been working for IBM since 1951 in other fields, so IBM wised up and began using his expertise!

Beanbag ashtray?!

It's hard to believe, but our honourable recent ancestors actually invented a 'beanbag ashtray' in the 1970s.

It was a combination of a stupid ashtray (don't smoke cigarettes!) glued to a tiny beanbag – Why? To sit level on uneven surfaces. This 'desirable' object would be a challenging test of a combined barcode/scanning system, because the barcode was glued onto the bumpy flexible bottom of the tiny beanbag.

The Symbol Selection Committee had pretty tough specifications. The barcode had to be less than 1.5 square inches (about 10 cm²) in area. It had to be printable with the then-current label-printing technology, readable from any direction and at different speeds, and have fewer than one error in every 20,000 scans.

The IBM team soon realised the problem was that perfect printing was impossible! A tiny smudge on a label made the bullseye barcode unreadable. So they said goodbye to the bullseye barcode, and opted for a rectangular barcode. George Laurer was the IBM engineer who actually designed the rectangular barcode.

So now the barcode lines could be printed along the length of the line, not across. And any smudging just made the line a little longer, not wider – it was still perfectly readable.

And did those tweaks make a difference!

By early 1973, IBM had built a scanning system that could read a barcode on the bottom of a (wait for it) beanbag ashtray (a typically kitchy 1970s product) as it went flying over the scanner after being hurled by a semi-professional softball pitcher.

The IBM 3660 incorporated scanners, computer terminals and a local area network connecting everything to a central computer. A red optical laser read the barcode, which was just a series of lines of varying widths, and worked out the distance between the leading edge of each line. Sound familiar? Yep, they had developed our current barcode-scanning system. IBM just had to hope that the Symbol Selection Committee would choose their rectangular barcode design over the other designs.

On 3 April 1973, the committee did just that: it gave IBM's rectangular barcode the go-ahead – the barcode design and barcode-scanning system we still have today.

UPC

The 'barcode' comes in a few hundred different varieties. But only a few dozen are widely used today. The UPC that supermarket shoppers might recognise is just one of these many varieties.

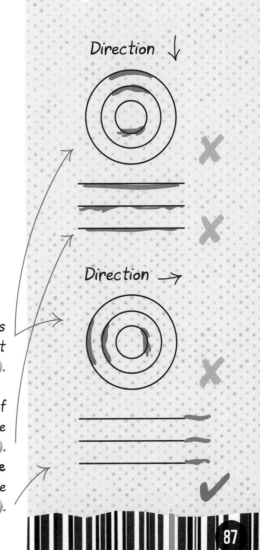

Direction ↓

Direction →

CIRCULAR BARCODE: If the ink bleeds or smudges a little, whichever way the label is printed, lines get thicker and make the barcode unreadable ☹.

RECTANGULAR BARCODE: If the direction of printing is **across** the bars of the code, ink smudge will make the lines thicker and unreadable ☹. However, if the direction of printing is **along the length of the bars**, ink smudge only lengthens the bars – **not** making them thicker and unreadable ☺.

THE FIRST MODERN BARCODE SCAN

Just over a year later, at 8.01 am on 26 June 1974, the first ever item with the modern barcode on it was scanned. The shop was the Marsh Supermarket in Troy, Ohio. And that very first item was a ten-pack of Wrigley's Juicy Fruit Chewing Gum, costing 67 cents. They chose something small and a little bumpy, to see if the barcode reader could handle it. The correct price popped up on the cash register on the first go.

The rectangular barcode worked flawlessly – both in recognising the grocery item, and in improving profits. It quickly became apparent that within five weeks of being installed in a shop, the dollar value of sales went up by about 10–12% and stayed there, because the processing of multiple grocery items at the checkout counter became so much faster. And in the case of Marsh Supermarket, the quicker and more accurate stocktaking dropped the shop's running costs by another 1–2%.

GRIM OUTLOOK FOR NEW-FANGLED INVENTION

Despite the barcode's universal prevalence in our lives today, it certainly was no overnight success. In fact, in 1976, *Business Week* published an article, 'The Supermarket Scanner that Failed', which painted a gloomy picture of the new invention's future.

Initially, there was resistance from the shoppers. They were being asked to buy a grocery product that did not have a price on each item, only on the shelf. How could they remember the price of each item as it zipped through the scanner if they had packed dozens of products into their trolley? What if they were being ripped off? Some stores provided grease pencils in the aisles, so shoppers could write the price on each item – but very few shoppers used these.

Sure, on the manufacturing side, by 1975, 75% of grocery produce had a barcode printed onto the label. But what good was that if the stores didn't have a scanner?

The new scanning machines were expensive.

The 'experts' had predicted that 1,000 stores in the USA would have scanning machines by 1976 – but the real number was only 50. By 1977, it was still fewer than 200.

But in the early 1980s, the mass merchandisers (such as Kmart) began to adopt the barcode scanners – only then did the barcode really take off.

The United Kingdom was a lot slower than the USA in adapting the barcode. By 1982, 70% of groceries in the UK came with barcodes.

Today, the use of barcodes saves over US$30 billion each year.

What started as a brainwave on a beach for Joe Woodland 25 years earlier was about to transform shopping and tracking worldwide for ever – as well as make it a whole lot noisier (with those beeps).

But those barcode-scanning beeps are really just white background noise. Imagine how much worse it could have been if Joe had stuck with his earlier invention – and had given us even more elevator music.

Not just any store

The night before that first test of the new rectangular barcode, a team of Marsh staff moved in and stuck thousands of adhesive barcode stickers on hundreds of items in the store. At the same time, the National Cash Register company, which was based in Ohio, moved their scanners and computers into the store.

The first 'shopper' was not just any random shopper. It was Clyde Dawson, the head of Research and Development for Marsh Supermarket. (He is pictured in the black-and-white photo opposite.)

ALKALINE DIET

Crazy diets are a sceptic's 'bread and butter'. The diet industry is itself a good example of crazy diets – over 95% of the time, advertised diets don't help with weight loss but, amazingly, the diet industry keeps on growing. So, what about the Alkaline Diet? Does it cut the mustard?

BODIES ARE ACE WITH ACID

Each day, our bodies generate about 10,000 to 20,000 mmol* of acids, mostly as carbon dioxide (as a result of normal metabolism, and generating energy from carbon-based foods). The exact amount varies, depending on how much you exercise. However, our bodies deal with all this acid quite easily. But what about food?

So how much acid does food generate, or carry inside it? Hardly anything!

For example, 100 g of beef generates about 8 mmol of acid, while a litre of Coca-Cola carries 2.5 mmol of acid.

So the amount of acid in our diet is inconsequential, compared with how much we generate naturally. We're looking at less than 100 mmol of acid from your diet versus 10,000 mmol from the activities of normal living.

*1 mole of any chemical substance is 6.02×10^{23} (Avogadro's number) molecules of that substance. So a millimole is one-thousandth of a mole.

90

HISTORY OF THE ALKALINE DIET

The Alkaline Diet popped up around 2002, courtesy of Robert O. Young with his series of books and videos. His first book, *The pH Miracle*, claimed that modern diets made our bodies too acidic (WRONG!), and this extra acidity was the cause of many diseases, such as cancer (WRONGER). Somehow, he shaped all his wacky ideas into an Alkaline Diet for weight loss. (However, *any* new diet will cause a temporary weight loss, so that's not much to crow about.)

The one element of truth in all this rubbish is that you can't be fit and well if your blood is too acidic.

But Young's claim that any diet could ever make our blood too acidic (or more alkaline, for that matter) is impossible. Diets can't change the blood pH in an otherwise healthy person. Furthermore, it is impossible for anybody to walk the streets, apparently well and healthy, and have acidic blood.

None of this stops the 'Alkalisers' one bit.

ACIDIC-ALKALINE 101

So anyways, what is this 'acidic–alkaline' thing?

Well, we use the pH scale to measure how acidic or alkaline a substance is. It's a weird scale that goes from 0 to 14, with 7 as the mid-point.

Right in the middle, a pH of 7 means that the substance is neutral – neither acidic nor alkaline (another word for 'alkaline' is 'basic'). Above 7 is alkaline, and below 7 is acidic.

As the pH drops right down towards zero, the stuff you're testing gets increasingly acidic. So tomato juice has a pH around 4, grapefruit juice is around 3, lemon juice has a pH around 2, while hydrochloric acid in the stomach has a pH varying between 1.5 and 3.5 (people usually say 'around 2').

As the pH climbs from 7 to 14, then the stuff is increasingly alkaline. Seawater is slightly alkaline with a pH around 8. Baking soda/sodium bicarbonate is more alkaline with a pH around 9, ammonia solution is around 11, while bleach and oven cleaner are very alkaline with a pH around 14.

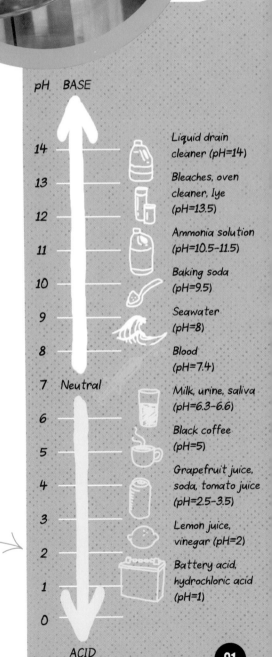

pH BASE

14 — Liquid drain cleaner (pH=14)

13 — Bleaches, oven cleaner, lye (pH=13.5)

12 — Ammonia solution (pH=10.5–11.5)

11 —

10 — Baking soda (pH=9.5)

9 — Seawater (pH=8)

8 — Blood (pH=7.4)

7 Neutral — Milk, urine, saliva (pH=6.3–6.6)

6 —

5 — Black coffee (pH=5)

4 — Grapefruit juice, soda, tomato juice (pH=2.5–3.5)

3 — Lemon juice, vinegar (pH=2)

2 — Battery acid, hydrochloric acid (pH=1)

1 —

0 —

ACID

HORSES &
THE ALKALINE 'MILKSHAKE'

It's not just humans that were lumbered with the Alkaline Diet – so were racehorses.

Horses run absolutely flat-out in a race. With their intense effort, they can't deliver enough oxygen to their muscles, so their muscles generate large quantities of lactic acid. And sometimes they get 'lactic acidosis', which adversely affects their performance.

Trainers, in an attempt to overcome this, invented the alkaline 'horse milkshake'. (And yes, it was an Australian invention, but it's not as delicious as you might hope. In fact, it's so bitter that the horses won't drink it.) A few hours before a big race, the trainer threads a tube into the stomach of the poor horse, and a mixture of sodium bicarbonate, sugar, electrolytes and water (and sometimes other 'substances') is poured in. The idea is that the alkaline sodium bicarbonate will reduce the level of lactic acidosis, and the horse will run the race of its life.

The horse milkshake study results are all shook up – the horses didn't win more races. But they did seem to be less fatigued after a race and to recover faster.

Today, horse milkshakes are banned in the USA and Australia. Even so, there are still some horse trainers who get busted for using them.

DIFFERING ACIDITY/ALKALINITY OF THE HUMAN BODY

Most of your body is alkaline, but some bits are acidic. However, the overall pH of your body is never acidic – unless you are very unwell. Your body is super-good at keeping itself in balance, regardless of the claims of the Alkaline Diet people.

Your overall pH is slightly alkaline, thanks to some very powerful mechanisms. These involve your lungs, your kidneys, and several chemical buffering systems.

Overall, your body is slightly alkaline. But there are various differences in pH throughout your body. And that's because your body has several 'compartments and organs' that are physically and chemically 'isolated' from one another.

The blood is slightly alkaline, with a pH of around 7.4 (it usually sits between 7.35 and 7.45). If the pH of your blood drops below 7.0 – in other words, if it becomes even slightly acidic – then you are either desperately ill in an Intensive Care Unit, or about to die, or both.

In fact, our bodies need a mixture of mostly alkaline areas, and a few acidic areas, to operate properly.

Saliva is slightly acidic. This helps an enzyme (amylase) contained in your saliva start the digestion process of food in your mouth. (Amylase breaks down indigestible long chains of carbohydrates – starch – into shorter chains that can be digested.) The salivary glands inside your mouth generate about 0.75–1.5 litres per day.

Try this very safe experiment. Roll a piece of non-crusty bread into a ball and mush it inside your mouth with your tongue, or just leave it on your tongue. Your salivary glands will quickly generate some saliva, and the amylase enzyme in it will start splitting the long, complex chains of carbohydrates (starch) into smaller chains of sugars. Suddenly, as you get a 'dissolving' sensation on your tongue, you will also taste a little sweetness in your mouth. (This also works with other foods, such as rice and potato, that are rich in starch but low in individual sugars, which are sometimes called simple carbohydrates.)

But your stomach is much more acidic than your mouth, with a pH around 2. Your stomach is so acidic that none of the foods you eat can effectively change its acidity.

Once your food has been mightily churned around inside your stomach, it then enters the small intestine. Surprisingly, the pH can

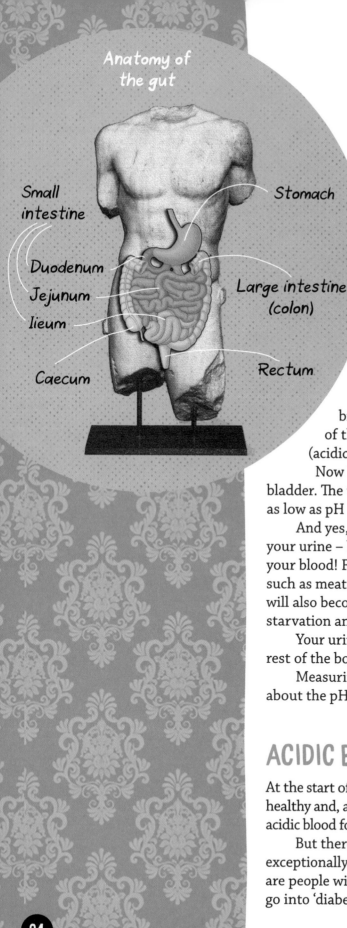

Anatomy of the gut

Small intestine

Stomach

Duodenum

Jejunum

Ileum

Large intestine (colon)

Caecum

Rectum

vary widely in different parts of the small intestine. (The small intestine is where the nutrients in food – fats, proteins, carbohydrates, etc – are absorbed into the blood supply that takes them to the liver.)

In the first part of the small intestine (called the duodenum) the pH is about 6 (acidic). (By the way, the duodenum gets its name because it's as long as the width, or breadth, of 12 fingers. 'Duo' is two, while 'denum' is 10.)

As the food progresses next through the jejunum (second part of the small intestine), and then the ileum (third and last part of the small intestine), the pH rises to about 7.4 (alkaline) by the end of the ileum. This change in pH in the ileum happens because the pancreas dumps alkaline bicarbonate ions into it. Once you get to the beginning of the large intestine (the caecum) the pH drops to 5.7 (acidic), but lower down rises to about 6.7 in the rectum.

Now let's leave the gut, and head to a different area – the bladder. The urine in your bladder is slightly acidic. The pH can drop as low as pH 5.5, but is usually around pH 6.2.

And yes, the food you eat can temporarily change the acidity of your urine – but that's completely different from changing the pH of your blood! Foods that acidify your urine include protein-rich foods such as meat, as well as cranberries, plum and prunes. Your urine will also become more acidic in uncontrolled diabetes, diarrhoea, starvation and dehydration, and a few respiratory conditions.

Your urine is neatly stored in your bladder, separated from the rest of the body, and on a one-way trip out, headed down the toilet.

Measuring your urine's pH with a dipstick tells you nothing about the pH of your blood.

ACIDIC BLOOD AND LIVE?

At the start of this story, I said that you can't walk the streets, well and healthy and, at the same time, have acidic blood. In fact, if you've got acidic blood for more than a few moments, you'll be lucky to survive.

But there are two groups of people who can recover exceptionally well from an acidic blood pH (i.e. less than 7.0). First are people with Type 1 Diabetes. If they run out of insulin, they can go into 'diabetic ketoacidosis' (in 'acidosis', 'acid' means 'acid', while

'osis' means 'being in a state of'). If their diabetic ketoacidosis is recognised and they quickly get the correct medical care (insulin), their recovery can be excellent.

The second group are newborns with respiratory (breathing) failure. They can't 'blow off' (remove) their carbon dioxide via the normal pathway of the lungs because they have difficulty in breathing. These newborns also can recover well when ventilation is provided.

Practically everybody who survives the very dangerous condition of acidic blood has been in the pointy high-tech end of modern medicine – the Intensive Care Unit (ICU).

Taking care of a patient in ICU is dangerous and expensive. The ICU staff are very skilled, highly trained and very knowledgeable. Sometimes, they can treat people with an acidic pH and bring them back to a safe alkaline pH – and into the Land of the Living.

The ICU team can infuse useful drugs, such as medications to make your heart work harder to push blood around your body better, and therefore stop your organs shutting down. They can totally control your breathing, by using a ventilation machine to shove air in and out of your lungs. They can bypass your gut, by infusing nutrients/food/vitamins/etc straight into the blood. They can take over (to some degree) the job of your kidneys with dialysis. And they do much, much more.

Patients who spend more than a week in ICU (especially if intubated and ventilated) take about a year to fully recover back to their original baseline.

But in most cases – if your blood/body pH goes acidic, and specialised medical care is not available, you will probably die. And a gulp of lemon juice won't fix you (read on).

DRINK ACID TO BECOME MORE ALKALINE?

The Alkaline Diet people claim that lemon juice (which is incredibly acidic with a pH of 2) will make your body more alkaline.

This is the exact opposite of common sense! It makes my brain hurt. It's like saying that if you're tired due to lack of sleep, then you can fix it by having even less sleep.

The irrelevant nugget of truth for this outrageous claim goes way back to when the Science of Dietetics was being invented. Early nutritionists were starting to feel their way into this huge and then-unexplored field. Along the way, they came up with the Acid-Ash Hypothesis.

ACID-ASH HYPOTHESIS

The so-called Acid-Ash Hypothesis is old! It was proposed over a century ago. (This was even before we had discovered the existence of 'vitamins'.) It has nothing to do with how the human body processes the foods we eat.

Your body digests your food slowly, over many hours, and via many chemical reactions. But according to the now-obsolete Acid-Ash method, you burn the food for a few seconds, with a simple flame.

The carbon in the food turns into carbon dioxide gas, the hydrogen turns into liquid water, and the nitrogen turns into various nitrogenous gases. And after the burning is complete, there is a tiny amount of ash left behind.

If the ash contains phosphorus, sulphur or organic acids, then the ash is slightly acidic. Foods that produce this type of ash are all meats, whole grains, seeds and beans.

But if the ash contains sodium, potassium, calcium and magnesium, then the ash will be slightly alkaline. The food groups that produce ash with these elements include fresh fruit and raw vegetables. So acidic lemon produces an alkaline ash.

Yep, it's totally true that these ashes can be acidic or alkaline. But this is irrelevant to how our bodies process food. The Acid-Ash Hypothesis was a scientific dead end.

Yet, bewilderingly, this is the theoretical basis of how the Alkaline Diet can supposedly cure you of almost any disease! Robert O. Young got into a lot of legal trouble for this outrageous claim – especially because people died as a result of his treatments.

ROBERT O. YOUNG AND $105 MILLION FINE

Back around 2002, Robert O. Young, an alternative medicine practitioner, turned the Acid-Ash Hypothesis into the Alkaline Diet fad. He falsely claimed that he had five degrees in science, including three 'doctorates' – a DSc, a PhD and an ND. (Three 'doctorates'! His lies were huge, not just little fibs.)

His website said that 'Dr Young is a scientist, health nutritionist, educator, and microbiologist'. He has also claimed that 'cancer is not a cell but a poisonous acidic liquid' (so very wrong, in so many ways). He also claimed that 'a tumor is the body's protective mechanism to encapsulate spoiled or poisoned cells from excess

CHANGE YOUR pH QUICKLY

There is one pathway that you can easily control to change your blood pH – breathing. Big difference: breathing is not eating – unless you're a Breatharian (in which case you reckon you can get all the nutrients you need to survive from air and sunlight).

To understand how breathing can change your pH, you need to know that you are always creating carbon dioxide from oxygen (about 27 g each hour). Also, your blood is about 55% salty water and about 45% cells (mostly red blood cells). Finally, if you dissolve carbon dioxide in water, you get a weak acid (carbonic acid).

Let's contemplate what would happen *if* we tried two experiments. (No, DO NOT DO THESE EXPERIMENTS YOURSELF; just ponder the effects they would have inside your body.)

Number one. Breathe in and out quickly and deeply for a few minutes. Breathing like this blows off way too much carbon dioxide out of your lungs. Because the carbon dioxide in your lungs has to be in balance with the carbon dioxide in your blood, the result is that you also remove carbon dioxide from your blood. The level of carbonic acid in the salty water in your blood will drop. This makes your blood more alkaline. (Yes, you can make your blood alkaline by breathing lots!) You might feel actual physical symptoms, such as tingling in your fingers, light-headedness, chest pain, and feeling like you will pass out and die. (See why you shouldn't actually do this experiment?) Once you change back to normal breathing, your breathing reflexes will take over, and you will recover.

Now for number two – the opposite. Stop breathing. This makes the carbon dioxide build up in your blood, and dissolve in the salty water that makes up about half of your blood. This changes your blood pH and shifts it towards the acidic end. (Again, don't do this – you're not meant to stop breathing!) As soon as you start breathing again, your blood pH rapidly shifts to being slightly alkaline again (hooray!).

$$CO_2 + H_2O \rightleftharpoons H_2CO_3$$

Carbon dioxide	Carbonic acid
In lungs	*In blood*

acid that has not been properly eliminated through urination, perspiration, defecation or respiration' (again, way out of the ball park, in so many ways).

Young squarely laid the blame of many conditions, especially cancer, on excess acidity in the body. What did he base these claims on? A broad and deep education? No.

In the 1970s, he enrolled at the University of Utah, and studied just one single course in biology. He never graduated. In 1995, he bought (yup, bought) several degrees from the unaccredited correspondence school Clayton College of Natural Health and got a PhD in less than a year. (At a real university, it usually takes three to five years of hard work to earn a PhD.)

He also set up a wellness centre, the pH Miracle Living Center, where people were charged US$2,000 a day for his treatments. In his Wellness Centre, he would wear a white coat and examine

Household cleaners – alkali vs acid

Up at the alkali end of the pH scale are a range of cleaners – chlorine bleach, oven cleaner, and tub and tile cleaner (pH 11–13), and ammonia (11–12). Alkalis are great at cutting through organics, such as oils, dirt, grease and proteins. But with great cutting power comes great potential damage. They can all damage fabrics, while chlorine bleach, ammonia and tub and tile cleaner give off nasty chemicals into the air, so you need good ventilation when using them. Baking soda (pH 8–9) is alkaline enough to remove dirt and grease, but is still fairly close to neutral (pH 7) to be pretty safe for most uses.

Regular soap is around pH 7–8. So, while it doesn't brutally and quickly remove all organic grimes, it's mild enough to be used for most cleaning, without having to worry about causing major side effects.

Heading towards the acid end, both lemon juice and vinegar are quite acidic at pH 2–3. Acids are usually better at removing rust, calcium and similar minerals. However, they can damage stone and grout.

Toilet bowl cleaner is highly acidic (pH 1–3). It's very effective but, again, you have to be cautious. And don't forget about needing especially good ventilation.

his patient's blood with a microscope (even though he had no qualifications in histology, microbiology or pathology). He also intravenously administered baking soda (baking soda, as used to make cakes and bread!) to at least six terminally ill patients.

One of them was the British army captain Naima Houder-Mohammed. She had been diagnosed with breast cancer. After paying Robert O. Young more than US$77,000 and receiving 33 sodium bicarbonate intravenous drips, she died at the age of 27.

His most infamous case was that of Kim Tinkham. She had been diagnosed with breast cancer and, in a TV interview with Oprah Winfrey, said that she would not have any surgery, chemotherapy or radiation treatment. She was 'treated' by Robert O. Young, and died.

In 2016, Robert O. Young was convicted for practising medicine without a licence. In November 2018, a jury awarded a US$105-million settlement against him, for a former patient of his, Dawn Kali. She had breast cancer, and he had advised her not to undergo conventional treatment but, instead, to take his 'pH injections'.

PEOPLE WHO NEED AN ALKALINE DIET

For most of us, eating sodium bicarbonate makes no difference to our health.

But in most 'myths', there is often a tiny nugget of Fact.

And yes, there are two uncommon conditions where a person's body can become dangerously acidic, and where having alkaline sodium bicarbonate (baking soda, or $NaHCO_3$) helps.

One is a very rare disease called Renal Tubular Acidosis, while the other is Chronic Renal Failure. In each case, the sufferers are prone to Metabolic Acidosis. And in each case, treating with sodium bicarbonate tablets will slow the progression of the disease.

Separately, it *seems* possible that ingesting 300mg/kg of sodium bicarbonate might help some athletes. Specifically, it might help athletes who do a single bout of supramaximal exercise (i.e. go at it as hard as they can, and then a bit more) or who do high-intensity intermittent activity. The problem is that the results from a bunch of studies are quite variable. Maybe it's a real effect, but maybe it's not – which makes it hard to say if it helps athletes push harder.

BENEFITS OF THE ALKALINE DIET?

The Alkaline Diet is definitely good for people who sell litmus paper (those little strips that change colour depending on the pH). That's because, on the Alkaline Diet, you need to use strips of litmus paper to measure the pH of your saliva and urine, and it's on those results that you're supposed to base the changes to your diet.

However, according to Clare Collins, Professor of Nutrition and Dietetics at the University of Newcastle (and an accredited practising dietitian, who also leads a large group of nutrition researchers at the university), claims of the general health benefits of the Alkaline Diet simply don't stand up.

People following the Alkaline Diet do tend to eat plenty of fruit and vegetables. That's excellent. But there's no point in testing your urine's acidity. (Urine pH has nothing to do with blood pH, remember?)

So, does the Alkaline Diet pass the Acid Test? Hell, no!

MORE REAL SCIENCE

Many meta-analyses have looked at alkaline diets. One summarised some 8,278 papers, looking for links between varying acidic-alkaline dietary intakes (and/or alkaline water) and any cancer outcome. Only one study was good enough to be counted as real science – and it found no link between acid load in the diet and bladder cancer.

Another meta-analysis looked to see if there was a relationship between foods that increased the acid load in your diet and osteoporosis – it found none.

ALKALINE WATER – SAME, SAME

Most regular drinking water out of the tap has a pH of around 6.5–7.5. But you can buy 'alkaline water' with a pH around 8.8 to 9.5. The crazy claims for alkaline water are that it can give you 'better hydration' and that it is 'designed to obtain optimum body balance'.

Plenty of us have never heard of alkaline water.

But in 2013, Beyoncé included alkaline water on the rider for her Mrs Carter Show World Tour. And she's a genuine Diva. So if it's good for her, then why wouldn't I want some too?

But, is there any science behind alkaline water? Nope.

Alkaline water fits into the rising 'connoisseurship' relating to plain old H_2O. In the USA, bottled water has now become the nation's most popular drink – even if it does cost more than petrol!

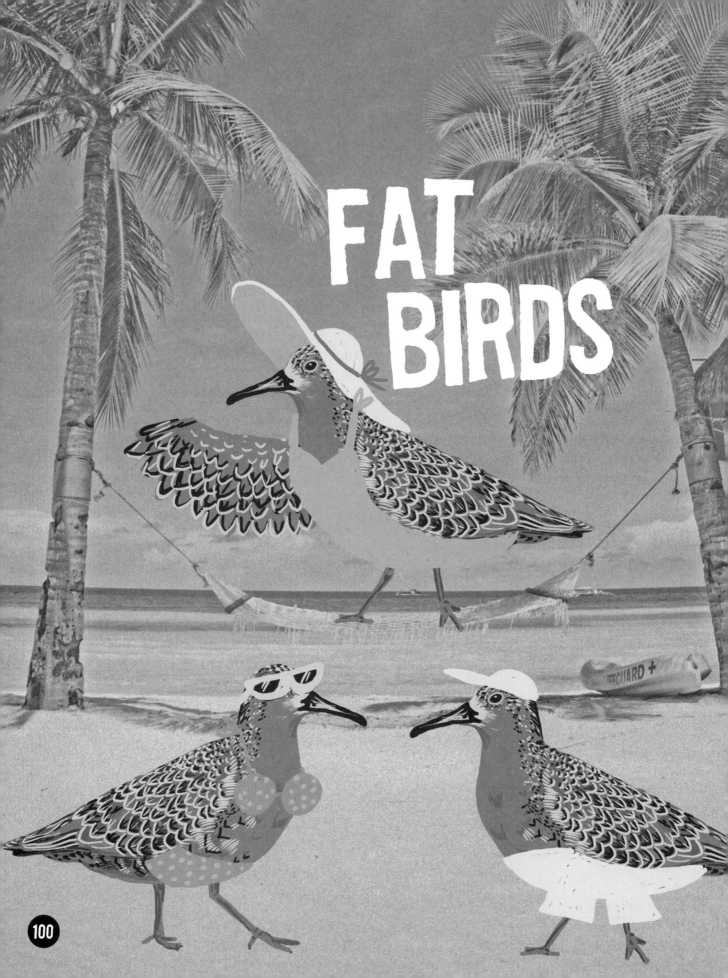

We humans are getting fatter. About 20 years ago, only 7-8% of Australians were obese, but nowadays it's up around 25%. In the USA, about 40% of the population is obese. It's been billed the Obesity Epidemic.

Now, there's an amazing bird called the bar-tailed godwit. So how is it related to people getting fatter?

Well, this is a bird that can double its weight in a week or so, and then burn it off again in the same amount of time. (But don't think that you'll be able to match its rapid weight loss ...)

FAT 101

Fat is a great way to store energy – dietetic scientists call it 'Nature's storehouse of energy'. Fat repels water – so you can store energy without having to store water, which is just dead weight. Weight for weight, fat stores about twice as much energy as carbohydrates and proteins.

In humans, most fat is stored as triglycerides. A triglyceride molecule looks like a capital 'E'. The vertical stroke is a glycerol molecule. The three horizontal strokes can be any of a bunch of different fatty acids, such as palmitic acid, oleic acid and linoleic acid. This big triglyceride molecule can very easily be broken down into fatty acids, which can then be used directly by the muscles as a fuel.

In vertebrates (animals with spines), about half the energy used by organs such as the heart, kidneys and liver comes from burning fat.

Why are athletes so scared of catching a cold?

This is a common question asked by the triple j Science Talkback audience around the time of each Olympic Games.

It turns out super-slim athletes are very vulnerable to infections, and that's because they generally have very little body fat. You see, some special immune system cells live in body fat. These immune system cells need instant access to lots of energy to quickly fight an infection.

Having low body fat means that these special immune system cells can't function. So, what might be a short-lived and mild common cold in you and me could leave an Olympic athlete bed-ridden and unable to perform.

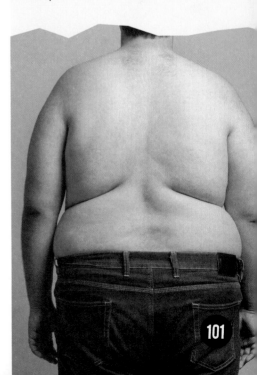

Bar-tailed godwit

So here's a little background on our star, the bar-tailed godwit. (We also talk about this plucky bird in **HOW DO BIRDS SLEEP?** on page 27.) It breeds on the Arctic coasts and tundra of Scandinavia, northern Asia and Alaska. Birds congregate in Alaska in July.

They fly south to cross the Equator and spend the northern winter in warmer climates, like New Zealand. They fly north again in their search for the eternal summer, and leave New Zealand in March.

There are about 1.1 million of them worldwide. They are about 37–41 cm long (measured bill to tail), with a wingspan of 70–80 cm. The males (190–400 g) are smaller than the females (260–630 g).

But in migrating birds and hibernating animals, fat supplies almost all the energy. Flying insects also get most of their energy from fat.

Fat is stored in special cells called 'adipocytes'. In humans, fat is stored mostly in the adipose tissue just under the skin. But there are also various deposits of fat scattered around the body – wrapped around the heart, in between the muscles and the small and large intestine, and so on. This fat is continually being broken down and then restocked. Even though your total amount of fat may stay the same, the atoms that make up this fat change.

The main advantage of carrying a bit of fat is that it will see you through lean times. The average American male carries enough fat (15 kg) to meet his energy needs for two months. A really obese person may carry 110 kg of fat, which is theoretically enough to supply their energy for a whole year!

The longest period of time for a person not eating is 55 weeks. This fasting man lost weight at the rate of 2.2 kg per week. He survived by burning up his body fat, drinking water, and taking vitamins and potassium supplements. He did not eat any food during that year and a bit.

FAT-FUELLED MIGRATION

When birds migrate long distances, they need lots of energy. They get the energy to keep on flying by burning up fat.

Your average non-migrating bird carries only 3–5% of its body weight as fat.

If you fly further, you need more fat. Birds that migrate short distances carry 20–30% fat. Some long-haul birds really ramp up their body fat, to 40–50%.

But our heavy hitter, the bar-tailed godwit, holds the record, at 55% body fat! This huge fat store is their secret weapon. It gives them enough energy to make them the long-distance flight

The flight paths of seven tracked bar-tailed godwits.

0   1000 2000

km

THE COLOUR OF FAT

The human body has different types of fat.

There's white fat (which stores energy as big, fat molecules). In humans, white fat above the waist is bad, in terms of health risk, but below the waist it poses less of a health risk.

If you get fatter above the waist, the existing fat cells just get bigger and plumper. But if you get fatter below the waist, you grow extra fat cells – an extra 1.8 billion fat cells for each kilogram of fat. Excess fat inside your belly is worse for your health than the loose flobbly type of fat just under your skin.

Brown fat is different from white fat. It's mostly in the upper body, often around the shoulder blades, and also near the kidneys and liver. When needed, it breaks itself down from big, fat molecules into a bunch of smaller molecules, generating a huge amount of heat at the same time. Babies burn brown fat to keep warm without shivering. If humans expose themselves to the cold, they start making brown fat. There's also some rather weak evidence linking brown fat to spontaneous human combustion (which I can't help mentioning because it's so fascinating, but which has never been proven – yet).

And finally, there's beige fat – a mixture of white and brown fat cells. (I don't know what it does, and I don't care – because I'm against beige anything on principle.)

The more we learn about fat, the more we learn how it's involved in energy metabolism.

champions – 11,680 km non-stop across the Pacific, in 8.1 days of continuous flight. This detailed flight data comes from actual measurements from surgically implanted 26-g battery-powered transmitters. Seven birds were monitored and flew non-stop for between six and nine days. This was a massive effort for these little creatures – they had to run their bodies at eight to ten times their normal metabolic rate, for all that time. They're covering a distance equal to the distance a jumbo jet can cover in one hop.

It is a massive and risky undertaking to fly so far over water in a single hop. So, what's the advantage?

Well, there are no avian predators in the Pacific capable of killing a migrating godwit. Their flight path is also free of pathogens and parasites. This is important, because long-distance endurance efforts, like this, suppress the birds' immune systems. And finally, New Zealand, their destination, has no predators of godwits.

But if they tried doing short hops along the coast, they might get eaten.

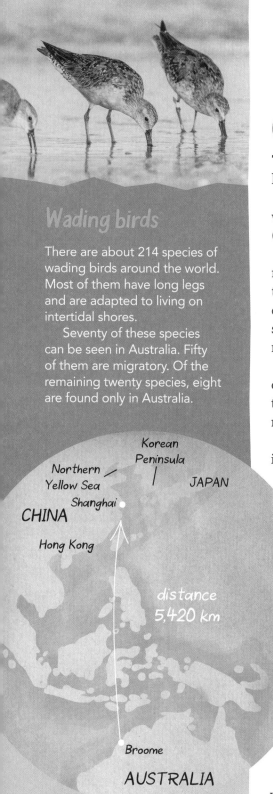

Wading birds

There are about 214 species of wading birds around the world. Most of them have long legs and are adapted to living on intertidal shores.

Seventy of these species can be seen in Australia. Fifty of them are migratory. Of the remaining twenty species, eight are found only in Australia.

Korean Peninsula

Northern Yellow Sea

JAPAN

Shanghai

CHINA

Hong Kong

distance 5,420 km

Broome

AUSTRALIA

This map shows that the birds depart from Broome and then fly to the Yangtze River estuary, in China, near Shanghai.

OTHER FAT BIRDS

The bar-tailed godwit is the long-haul champion, but other birds fly long distances too.

These include some 50 different species of Arctic migratory wader birds. Each year they travel some 12,000 km in search of food (but not in a single hop).

Broome in Western Australia is an essential stopover for migrating birds. It has five quite distinct habitats – each appeals to different birds. The waders prefer the enormous mudflats that exist between high and low tide. On the other hand, some 250 other species of bird love the other habitats – scrub, freshwater lakes, marshes and open grassy plains.

Wader birds that breed in the eastern Siberian mountains fly out so they land in Broome by March. That's perfect timing, because that's when the insects start appearing. The birds then move into a military-industrial pig-out mode.

Each bird will gobble up thousands of hard and soft invertebrates in a day. (Invertebrates are creatures that don't have a spine, such as insects that fly, or worms that wriggle in the mud.) The birds are so influenced by the tides that they arrange their day into two 12-hour cycles, rather than one 24-hour cycle.

The waders eat for eight hours and then rest for four, and then repeat. They'll spend 16 out of every 24 hours eating. The waders literally double their body weight, adding the extra mass as fat.

When the weather is good, usually around April and May, they hit the high road – on what the bird scientists call the East Asian-Australasian Flyway. There are about 3 million wading birds in Australia around this time; 2 million of them fly out of Broome. Smaller numbers fly out from the Gulf of Carpanteria and from Corner Inlet and Wilson's Promontory in Victoria.

They'll fly for three days continuously. They cover some 5,400 km, and all that hard-won fat just melts off as fuel. They're heading for eastern Asia, especially eastern China and Korea. They will rest and refuel. Again, they'll eat enough to double their weight, and then fly another leg and lose that weight.

It takes them until early June to arrive in Siberia. When you've got millions of birds in one place, well, they get down to some heavy petting and serious breeding.

SIBERIA – LUXURY LIVING?

So why would these wading birds head up to Siberia, which doesn't have a major reputation as a five-star luxury destination.

Well, it might not be luxurious for us humans, but for the waders, summer in the Artic Circle is Meals On Wheels.

In June, the air is thick with clouds of millions of insects. The newly born wader chicks have very short beaks, so they can't dig in the mud for the adults' regular meal of wriggling invertebrates. But the air is so full of flying invertebrates that virtually all they have to do is open their beaks and wait for the insects to fly in.

Sure, it's a very hard and long trip to get to Siberia, but it must be worth it because the birds do it every year.

After a few months in the Arctic Circle, the adults head back from Siberia to Japan, then over New Guinea and Cape York to New Zealand, and then cut across the bottom of Australia, turning right after the Great Australian Bight, before they end up at Broome again. The total trip is around 12,000 km. In their lives, these birds will cover a distance equivalent to flying to the Moon and back.

The modern human fat cat jet set has nothing on these fat birds, which have refined a lifestyle for themselves based on nothing more than flying, feeding and fornicating.

PREGNANCY DOUBLE WHAMMY

Can a woman get pregnant again if she is already pregnant? In other words, is it possible for her to have two babies of different ages in her uterus?

The technical term for this is 'superfetation' (also spelled 'superfoetation'). In humans, this phenomenon is possible – but it's very, very uncommon.

In fact, there is scepticism about whether it happens at all. We have evidence that is pretty good, but it's still not yet rock solid.

TWIN PREGNANCIES?

The great Greek thinker Aristotle wondered about superfoetation about 2,500 years ago. He was thinking about how very different a pregnancy could be in the hare (a bit like a rabbit, with very long hind legs) and the human.

He had noticed that when a hare gave birth, quite often the bunnies in a single litter would fall into two quite different sizes. There would be a bunch of full-sized, robust baby hares. And often there would be another bunch of scrawny, frail baby hares.

He thought it was likely that the mother hare had become pregnant, and then while already pregnant, had conceived a second time. There was nothing fundamentally wrong with the smaller weedy bunnies, Aristotle thought. Their only problem was that they had not spent enough time growing in their mother's uterus. And he turned out to be right.

It happens in a few other species as well (but, Aristotle thought, not in humans).

About 1–2% of all pregnant cows show signs of being potentially fertile by ovulating eggs – even though they are actually already pregnant. 'Superfetation' is said to happen in animals such as the mouse, the rabbit, the sheep, the buffalo, the cow, the burro, the mink, the roe deer, the rat, the cat, and the Australian swamp wallaby and kangaroo.

DETOUR

Superfoetation vs superfecundation

Superfoetation is different from superfecundation. The words sound quite similar – but there's a world of difference between them.

Superfetation happens when two eggs (ova) are fertilised at different times – days, weeks or even months apart. So there has to be a time delay between the eggs being fertilised.

In superfecundation, a female ovulates two or more eggs, in the same oestrus cycle, at roughly the same time. The two different eggs are then fertilised by the sperm of one or more males. This can lead to 'twins'.

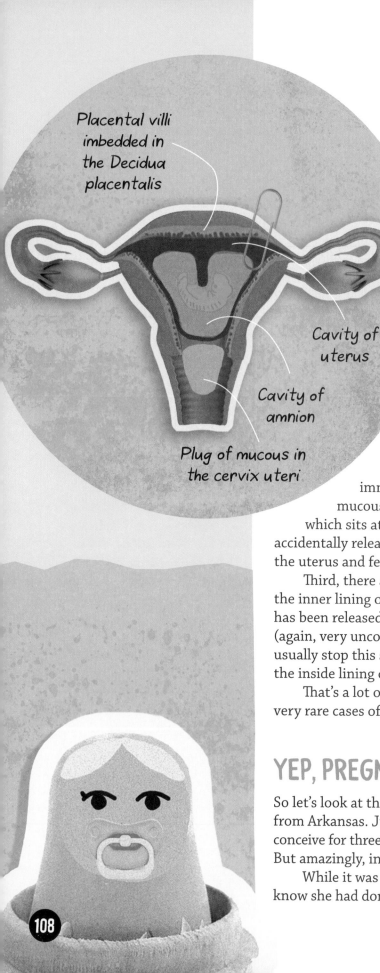

Placental villi imbedded in the Decidua placentalis

Cavity of uterus

Cavity of amnion

Plug of mucous in the cervix uteri

PREGNANCY 101

Superfetation is extremely uncommon in *humans*, with fewer than a dozen cases in the medical literature.

We humans usually have three separate 'barriers' to prevent a woman from getting pregnant again (once she is already pregnant). To summarise, the protective changes in pregnant women include: preventing another egg being made, preventing sperm from getting to another egg, and preventing a fertilised egg from implanting.

But let me go into more detail.

The first barrier involves the release of hormones, which stops the ovaries from releasing another egg.

The second barrier, which happens almost immediately after conception, is the formation of a mucous plug in the cervix (the opening of the uterus, which sits at the top of the vagina). Even if a second egg is accidentally released, this plug stops any other sperm from entering the uterus and fertilising that second egg.

Third, there are physical and chemical changes that happen to the inner lining of the uterus in pregnancy. So even if another egg has been released (very uncommon), and then fertilised by sperm (again, very uncommon), then these chemical and physical changes usually stop this second fertilised egg from being able to attach to the inside lining of the uterus.

That's a lot of hurdles to jump. Even so, there have been a few very rare cases of superfetation in humans.

YEP, PREGNANT WHILE PREGNANT

So let's look at the case of Julia Grovenburg, a 31-year-old woman from Arkansas. Julia and her husband, Todd, had been trying to conceive for three years, and had begun to consider adopting a child. But amazingly, in 2009, Julia finally became pregnant.

While it was clear to Julia that she had conceived, she didn't know she had done so *twice*, about two-and-a-half weeks apart.

She already had taken hundreds of home pregnancy tests over the years, but they had all been negative. Now that she had a positive result, she made a doctor's appointment. The ultrasound showed two independent babies, each living in their own separate amniotic sac. That's not too odd – people do have twins.

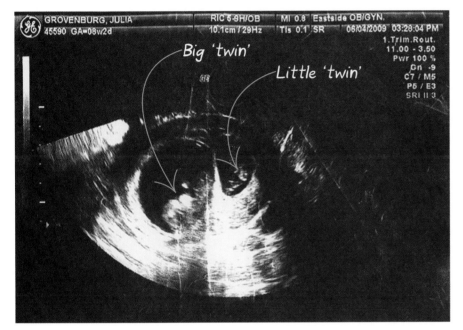

Proof in hares

It's definitely been proven that superfoetation happens in female European brown hares. It took some 2,500 years after Aristotle to develop the technology to show that superfoetation was real.

One robust study involved ultrasonic examinations of pregnant European brown hares from Day 3 of gestation. Sure enough, ultrasonography in several of the hares showed the appearance, at Day 11 of gestation, of a second live embryo inside its own amniotic sac.

However, in Julia's case, one baby was much larger and more developed than the other. This *was* odd.

There are medical conditions in which twins who are conceived at the same time can grow at different rates. But that wasn't happening here.

As the ultrasound scans mounted up for Mrs Grovenburg, both foetuses seemed to be growing normally and healthily – but always with one foetus consistently bigger than the other. It appeared as though a second foetus had simply arrived in the uterus and started growing two or three weeks after the first foetus.

The evidence certainly seemed to indicate that Mrs Grovenburg had conceived a second baby a few weeks after becoming pregnant with the first baby. And, yes, the happy ending is that both babies (Jillian and Hudson) were born on the same day, alive and well, by Caesarean section.

Cats superfoetate too!

Differential growth in multiple pregnancy?

Why is there some scepticism about the possibility of superfoetation in humans? Well, it turns out that under certain conditions, foetuses in multiple pregnancies can sometimes grow at different rates.

It might be that the placenta simply cannot fully support more than one foetus.

Or sometimes, the foetal blood can be transferred from one twin to the other. This is called Twin-to-Twin Transfusion Syndrome (TTTS). This is potentially lethal to both twins.

Or one of the babies having an infection can be another cause of them growing at different rates inside the uterus.

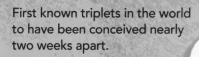

First known triplets in the world to have been conceived nearly two weeks apart.

PROVING SUPERFOETATION?

Although it seems very likely that Mrs Grovenburg conceived twice, a few weeks apart, we don't have rock-solid proof. For that, we would need a series of ultrasound scans taken right from the very beginning.

Ultrasound scans done up to the twelfth week of pregnancy are extremely accurate at determining an embryo's gestational age. This is because all embryos grow and develop at very close to the same rate.

If you had daily ultrasound scans from Day 1, you would see a series of changes. Initially, there would be no foetus or amniotic sac in the uterus. A while later, a single foetus in its single amniotic sac would be visible. And then the proof – after a few weeks, a second foetus would suddenly appear in its own second amniotic sac. Finally, both foetuses would develop normally, and the difference in size between them would be maintained as they continued to grow.

But most people don't usually get a series of ultrasound scans immediately after sex, so this hard data is difficult to get.

IS THERE ANY BENEFIT TO A DOUBLE PREGNANCY?

There is almost certainly a biological benefit to superfoetation. The evolutionary advantage is that females can have about one-third more offspring in each breeding season.

But for humans, more babies each breeding cycle isn't always a plus – especially if you're trying to get a good night's sleep!

PREGNANCY FACTOIDS

The 'duration' of pregnancy is a little 'odd'. The magic number most people know is '40' weeks. And yes, on average, the child is born at 40 weeks – however, this is 40 weeks from the start of the Last Menstrual Period (LMP). But ovulation is typically two weeks after the last period started. This means that, strictly speaking, the baby was born 38 weeks after fertilisation. So the baby is actually 38 weeks old at birth, not 40 weeks.

In 2012, there were about 213 million pregnancies worldwide. The vast majority (190 million, 89%) were in the poorer countries, while the minority (23 million, 11%) were in the wealthier countries.

There are lots of minor pregnancy side effects for women to put up with. One very common complaint is that pregnant women have to wee much more often. There are a few reasons for this. The pregnant woman has more circulating blood (because she now has an extra load to nourish – the embryo), so the kidneys extract more urine from this extra fluid volume. Also, the kidneys switch into a mode where they filter more urine from each litre of blood. And finally, the baby presses on the mum's bladder – and sometimes even kicks it!

This adds up to more wee for the mum-to-be.

Sometimes, the pregnant woman gets a gum inflammation – called gingivitis. The gums become more red and swollen, and can sometimes bleed. This is due to higher levels of oestrogen and progesterone during pregnancy.

And, baby, that's just the beginning …

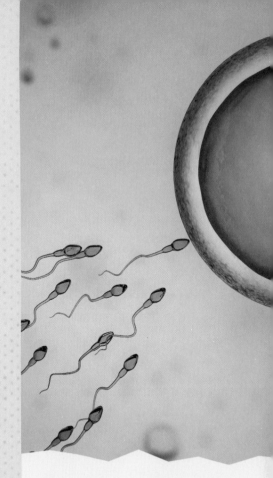

Human superfoetation

In 2007, in the UK, there were two separate cases of women conceiving a second baby three weeks after the first. There were no complications at birth, because the babies were each close enough to full term when they were delivered.

Two other cases on record involve the second conception happening one month later, and two months later.

One case in Brisbane in 2015 involved two baby girls, conceived ten days apart. The big surprise was that the couple had sexual intercourse just once, before the two conceptions! That implies that the sperm were still able to fertilise another egg, ten days after ejaculation (this is a bit longer than what we currently think sperm live for).

FIRE IN THE SNOW –
INCINERATING TOILET

Japanese hotel and public toilets have a high 'Wow!' factor.

Many have inbuilt water sprays that can be adjusted in so many exciting ways – *where* they spray, *how hard* they spray, the *temperature*, and so much more. I thought they were the best toilets in the whole world.

But that was before I went to the Land of Ice and Snow – Antarctica. There, for the very first time, I ran into a toilet feature I'd never seen before. Not even in Japan. Fire!

Australia has a science research base in Antarctica at Casey Station, on the coast. They always operate with the goal to create as little waste as possible. And this includes the toilets.

Enter the INCINOLET Electric Incinerating Toilet (Model TR) – or, as I like to call it, the Fire Breathing Dragon.

The Incinerating Toilet was at Wilkins airstrip, some 70 km inland of Casey Station. Of course, once I'd found out that a fire-breathing toilet was an actual thing, I had to try it out.

First, I looked into the bowl, and was filled with joy. The bowl was made of stainless steel, one of my most favourite materials. It was hinged in a clamshell design.

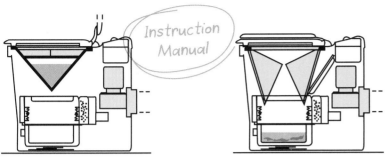

Instruction Manual

1. Drop bowl liner into toilet bowl. Bowl liner catches and contains all waste plus paper.

2. Flush bowl by stepping on foot pedal. *See side bar on this page.

3. Push start button to incinerate waste automatically.

Then, when I pressed my foot on the pedal, the clamshell opened up. I could see the remains of another person's No. 2 gently glowing, like the embers of a friendly backyard barbecue. There was definitely no nasty smell – hooray! The Model TR takes 90–105 minutes to turn a freshly deposited No. 2 into clean, germ-free ash. Each 'cycle' uses about a kilowatt-hour of electricity.

The Incinerating Toilet comes with instructions on the lid – which is lucky! I first had to do my No. 1 in a 'separate place'. Then I placed a paper bowl liner in the bowl, and did my No. 2 into it – like I was supposed to.

The next instruction said, 'FACE TOILET TO FLUSH'. The only way I could do that was to finish ALL my 'No. 2 business', then stand up, turn around and press the pedal. The word 'flush' had nothing to do with running water. It just meant that 'stuff' would 'go away' and vanish into the incinerator pan below. Finally, once the stainless-steel clamshell was closed, I was supposed to press the start button to fire up the toilet.

But I wanted to see behind the scenes (typical scientist!). I gently pressed on the pedal at the same time as pressing the start button. A burst of incineration came up the toilet bowl! I immediately let go of the pedal.

Seeing the sudden glow deeply impressed upon me the need to *always follow instructions as directed*.

I mean, how would you explain a burnt bottom in icy Antarctica? After all, Antarctica is not supposed to be a Flash in the Pan …

*Foot pedals and hand handles?

I later read all the instructions. The manual kept referring to a 'foot pedal'. Being a bit of a Grammar Vigilante, I immediately got annoyed.

The Latin root 'pes, pedis' refers to 'foot'. So, the lever that you operate with your foot is called a pedal. A 'pedal' is already something you operate with your foot. Calling it a 'foot pedal' is like calling it a 'foot foot pedal'. After all, the lever that you operate with your hand is called a 'handle'. You don't call it a 'hand handle'.

At the very least, please call it a 'foot lever'.

But I have learnt with the years to be less concrete and more kind.

SPAGHETTI SNAPPING

At the 2012 Olympic Games in London, the Cuban pole vaulter Lázaro Borges began his sprint towards the bar, holding his pole.

He was trying to clear 5.35 m. As he had done a thousand times before, when he got to the 'right' distance from the bar, he planted the far end of the pole into the ground.

The pole dug in and began to bend as it absorbed some of the energy of his forward run. It then began to straighten out, and started lifting him into the air. Suddenly, with absolutely no warning, when his feet were about 2 m off the ground, his pole snapped – into three pieces.

Luckily, he wasn't injured – but why did the pole break into three pieces, not two, or four?

FROM LITTLE THINGS, BIG THINGS GROW

This is not a simple problem to solve. One of the greatest physics minds of the 20th century, the Nobel Prize winner in Physics Richard Feynman, got very tangled up in it.

Back in the mid-1980s, two men simply wanted to cook a spaghetti dinner. They were Richard Feynman and his friend W. Daniel Hillis, a supercomputer innovator.

The problem was that the sticks of spaghetti were too long for the pot. Easy – just snap the sticks in half, into two smaller pieces. The spaghetti would be short enough to fit into the pot, and still be long enough for them to wind up on their forks.

Easy to say, hard to do.

No matter what different method they tried to break the uncooked spaghetti so it would fit into the pot, it wouldn't neatly snap into two equal lengths. To their frustration, it kept on breaking into three or more fragments.

Hillis said, 'We ended up, at the end of a couple of hours, with broken spaghetti all over the kitchen and no real good theory about why spaghetti breaks in three.'

Right there is one problem with being a physicist. Trying to understand the world around you can interfere with the simplest things – even cooking your dinner!

Furthermore, why bother with such a silly problem? It turns out that it's not so silly.

SIMPLE PROBLEMS CAN BE HARD

Another Nobel Prize Winner in Physics, Pierre-Gilles de Gennes, also tried to solve the problem of why spaghetti doesn't break into two pieces when you bend it.

In 1991, on French television, he said that the 'spaghetti mystery' was both very simple, and totally unsolved.

BREAKING UP IS NOT HARD TO DO

And, yes, 'fragmentation' is another field of knowledge that appears virtually everywhere in the natural and artificial worlds. 'Fragmentation' is the process by which any object shatters or breaks or splits – not just rods.

In rod we trust

Rods (of which spaghetti is just one example) are everywhere in the natural and artificial world around us.

There are rod-like structures in trees, in semi-flexible polymers, in carbon nanotubes, in active liquid crystals, and even in telephone networks and the Internet. We find rods in bones, the steel struts of a skyscraper, the legs of Water Strider spiders, fibroblast cells that are involved in repairing wounds and making connective tissue, the tiny 'electric motors' that rotate the flagellae of small bacteria, and even the cytoskeleton inside the axons of our nerve cells.

So understanding what happens to rods when you put a lot of stress on them is much more than just an interesting problem in mechanical engineering. Rods are basic to so many structures.

All rod-like structures can fail, when the stress is large enough.

When they fail, they often break (or fragment) into many segments of different sizes. Sometimes the results are catastrophic. That's why we've been studying how rods crack and snap for over a century.

Dry pasta
(1.9 mm diameter, 24 cm long,
one-quarter of a thousandth
of a second between pictures)

And, you guessed it, we still don't fully understand it and, yes, it's really important.

We see fragmentation happening at the big end of the scale – in the collisions of asteroids, comets, planets and neutron stars.

We see it on the small scale when subatomic particles collide.

But fragmentation also happens in the middle scale – in the smashing of windows, the exploding of bombs, and the pulverising of mineral ores. It happens in brittle structures such as the blades inside car turbochargers and jet engines, the cutting bits used to machine high-strength alloys, and in body armour.

And yes, fragmentation is an essential part of the relatively harmless breaking of spaghetti.

FRAGMENTATION 101 – BRITTLE

Engineers and scientists started doing the latest round of experiments into fragmentation in the early 1990s. They wanted to see what happened when materials got so overloaded with stress that they failed.

They put brittle glass rods into strong steel cases, and then simply dropped the steel cases from varying heights onto hard concrete floors, and counted the number of fragments created.

Depending on the height of the fall, they found different numbers of small and large fragments. The scientists came up with a few mathematical descriptions of what was happening.

But they needed more information. Surprisingly, this came from flexible materials.

FRAGMENTATION 102 – BRITTLE AND FLEXIBLE

The next series of experiments were done with both brittle rods (such as pasta or glass) and flexible rods (such as steel and Teflon). The pasta included San Giorgio #8 spaghetti and Barilla angel hair. The scientists stood the rods vertically on end, and then smacked them with pieces of metal moving at speeds up to 100 km/h. They filmed

the impact with a high-speed video camera at speeds of up to 62,000 frames per second.

In the first stage of the impact, both the flexible and the brittle rods began to buckle at regular distances along their length, taking on a wavy, curvy shape.

In the second stage of the impact, the flexible steel and Teflon simply became more 'wavy' or sinuous.

But the brittle materials did not get more rippled or wavy – they shattered. And they tended to shatter at rather specific lengths.

The points at which they broke up were the peaks and valleys in the wavy pattern running through the spaghetti rods just before they shattered. That's a very interesting clue …

FRAGMENTATION 103 – TWANG

The third series of experiments in 2005 tried to replicate what happens in the kitchen. If the spaghetti is too long to fit into the pot, you grab each end with a thumb and forefinger, and then simply bend the spaghetti until it breaks.

The scientists' version of this kitchen method was to clamp one end of the spaghetti, and then bend the other end until it almost (but not quite) broke. But then they let go of the bent end. They expected that it would just twang back and forth, like a diving board at a swimming pool after the diver has leapt off.

But the images from the high-speed camera surprised them. When the rod of spaghetti was released from being bent, little ripples raced along the rod to where it was clamped, and then reflected back. When the crests of the incoming and reflected waves momentarily passed over each other, the little waves tripled in size into a really big wave. The brittle spaghetti couldn't handle the stress, and snapped at this one point – which was hardly ever in the centre of the rod. (*That's* why it's so hard to break spaghetti in half!)

The first micro-crack in the wall of spaghetti appeared relatively slowly. It took about ten-thousandths of a second to appear. But the next phase, for the crack to completely open up, was much quicker. It took only ten-millionths of a second for the rod to completely snap.

But this snapping then released more energy in the form of ripples, again causing more breaks. This cascade continued until the spaghetti had broken into multiple fragments – often up to ten of them.

Back in your kitchen, that's why spaghetti won't break into only two parts. When you bend the spaghetti, you load up the rod with energy. You can't see them, but you are creating tiny waves that race up and down the spaghetti. These waves are quite short in length, so there are lots of them along the rod. Soon there are several 'peaks' of piled-up energy in just one spot – and bingo, that's where the spaghetti breaks first.

The vibrations travel into the remaining segments on each side of the first break – which then snap. And the cascade of breakages continues, on and on.

We've been cooking spaghetti for hundreds of years, but it wasn't until 2005 – when the results of these experiments were published – that we solved the problem of why a stick of spaghetti won't break neatly into just two equal fragments when you bend it.

TWIST AND QUENCH

So even though you know that breaking spaghetti in half is hard, what if you still desperately want to do it?

Well, that snappy knowledge took another 13 years to pry out of Nature – just in 2018. The scientists summed it up as 'Twist and Quench'.

Here's what to do.

Start with a typical 25-cm-long single strand of stiff and brittle spaghetti.

Grab one end with your thumb and forefinger, and hold it fixed.

Then, grab the other end with your other thumb and forefinger and then – and here's the key – *twist* the spaghetti! You are now shoving a different type of energy into your rod of spaghetti. It's not 'bending' energy, it's 'twisting' energy.

How far do you twist? At least 250° – about three-quarters of a full circle. It takes a fair effort to do that to a stick of spaghetti. But here comes the exciting part!

Now bend the pre-twisted spaghetti. Because it's already loaded up with energy, it won't bend as far before it breaks.

And remember the little rippling waves of energy that run up and down the rod of spaghetti? Well, because you haven't bent the spaghetti so far, these waves probably don't carry enough energy to set off multiple fractures.

So finally, your spaghetti will break into just two equal parts. Hoorah!

As long as …

You can see the next problem.

Yes, the pasta will break into two equal lengths. But there are hundreds of spaghetti rods in your spag bol. How much time will it take you to individually twist each rod of spaghetti to at least 250° before you bend it?

Perhaps a truly ingenious engineer will come up with a kitchen appliance to do the job – and maybe throw in a set of steak knives that will never go blunt.

Sounds like something that would cost a pretty penne.

Now, was that pun a little too saucy?

Pasta la vista, baby!

DR KARL'S Q+A

Marco Polo & spaghetti

A triple j listener rang in with a 'settle-a-bet' question: 'Did Marco Polo bring spaghetti back from China to Italy?'

The Chinese have had pasta for ages. Archaeologists have found a 4,000-year-old overturned bowl of tangled pasta.

The story of Marco Polo bringing back spaghetti was invented in 1929, in the Macaroni Journal – an official publication of the National Macaroni Manufacturers Association of the United States (now known as the National Pasta Association). It claims that a sailor, named Spaghetti, came across a Chinese couple mixing some dough. Some of it had spilt onto the ground, and dried. It could be easily reconstituted by boiling in water. But 'spaghetti' means 'thin string', so this sounds very unlikely.

Furthermore, when Marco Polo returned from China around 1295, pasta had already existed in the Mediterranean for some 2,000 years.

Let's twist again

119

VAPING &
E-CIGARETTES –
THE OLD
BAIT-AND-SWITCH

DETOUR

If you want to name a bad product doing bad things, then tobacco is a perfect place to start.

In the USA today, tobacco is the leading cause of preventable death – some 480,000 people die each year from cigarette smoking and from second-hand smoke exposure. Heartbreak from losing a loved one is hard to put a price on; but the health costs of tobacco-related illness and deaths add up to some US$170 billion each year.

And it's the community that pays this cost – not Big Tobacco. In economic terms, shifting the cost to someone else (who did not choose to carry that cost) is called a 'hidden externality'.

Given all the public health campaigns about the risks of smoking, how has Big Tobacco responded? Well, it's using an old sales trick called 'bait-and-switch'.

And the bait is the e-cigarette.

E-CIGARETTES 101

An electronic cigarette or e-cigarette is a small hand-held, battery-operated device that heats a liquid to produce a vapour that users inhale. (E-cigarettes are also called 'vape pens', and the experience is called 'vaping', from the 'vapour'.) The e-liquid usually (but not always) contains nicotine. The device gives the mouth a sense of smoking an actual tobacco cigarette.

The concept goes back to 1963, but the modern version was invented in 2003 by Hon Lik, a Chinese pharmacist whose father died from lung cancer (from smoking). Today's nicotine-based e-cigarette is minus the tobacco, but the addictive nicotine is usually there. The user breathes in the vapour and gets their nicotine hit while avoiding the 100+ known carcinogenic chemicals in tobacco smoke.

So are e-cigarettes less harmful than regular cigarettes? Probably yes. But are e-cigarettes totally safe? Definitely no. It's early days for e-cigarettes, and there is much we don't know about them.

In the e-cigarettes that contain nicotine, the addictive drug is carried into the lungs by some kind of flavoured vapour. Some of these chemicals used as carriers for the nicotine vapour have been approved for use as food additives, but not as substances to be

Speedway to death

To make trillions of cigarettes each year needs a lot of high tech. Today, a typical cigarette machine can make 20,000 cigarettes each minute. Just think about all the cigarettes we humans smoke in each second. If you laid them end to end, they would reach a length of 10 km. So here comes probably the weirdest statistic of the week: we humans smoke about 10 km of cigarettes each second, which is faster than satellites orbit the Earth.

Each individual cigarette stick generates a profit of about one single cent for the tobacco company. So, for each million cigarettes sold, Big Tobacco makes about $10,000, and one human dies. Rather unfairly, the medical costs of that person dying are much greater than $10,000.

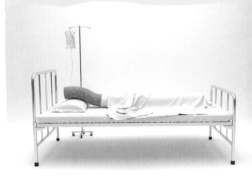

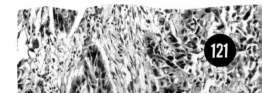

inhaled into the lungs. E-liquids can also contain other dangerous ingredients, including insecticides.

E-cigarettes don't come with enough health warnings, and still carry the risk of nicotine poisoning, especially for little kids. What's more, the containers in which they're packaged are not always child-resistant – some companies even market their products so that they look like lolly packets. At least one infant has died in Australia from swallowing a concentrated nicotine solution.

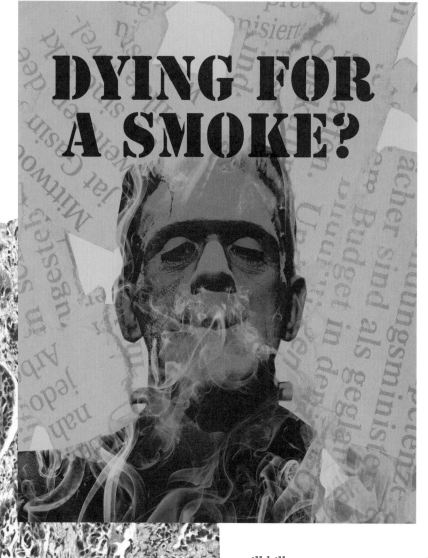

GET THE NEXT GENERATION

Big Tobacco has a long history of lying about the dangers of smoking cigarettes. Deceit combined with very clever marketing has been very powerful.

Cigarette advertising used to promote cigarettes as a way to define yourself – smoking could be an act of feminism or masculinity, or an easy way to be cool, or even the stamp of an international traveller.

Actually, even if the advertising was really awful, cigarettes would still sell, because the greatest hidden advantage that the tobacco companies enjoy is that nicotine is as addictive as cocaine and heroin.

Over the last few decades, there has been a massive worldwide pushback by governments against Big Tobacco. Certainly in the Western countries, there has been a huge drop in the numbers of teenagers taking up smoking.

So Big Tobacco responded. After all, when you know that your product will kill your current users early, then you definitely need to fish for new customers.

Enter the 'bait-and-switch'. You know how it works. You walk into the shop to buy one thing, but a persuasive salesperson talks you into an upgrade, or extras.

American tobacco giant Altria (previously known as Philip Morris) are using e-cigarettes as bait for their primary product. They have invested $12.8 billion into the company Juul, which makes the most popular e-cigarettes used by American teenagers. In fact, 'Juuling' is the term to describe using a Juul e-cigarette.

In the stores, Altria deliberately displays both the Juul vaping products and their regular cigarettes side-by-side. This is Big Tobacco's version of 'Bait-and-Switch'.

AND THE BEAT GOES ON …

Despite the steady decline of smoking in Australia and the USA over the past few decades, especially among young people, teenagers now are taking up vaping in growing numbers – because they can be cheaper than traditional cigarettes, they're seen as novel and cool, and are supposedly safer than smoking. The US market for vaping is US$5.5 billion (2018). In Australia, it was AU$50 million (2017).

Big Tobacco knows the teenagers are using vaping as a step towards smoking regular cigarettes. (Yep, it's what's known as a 'pathway' or 'gateway' drug.)

In the USA, in 2017, the percentage of school students who had tried e-cigarettes was 13% for 8th-graders, 24% for 10th-graders, and 28% for 12th-graders.

So Big Tobacco is still harming the next generation. And just like in the old days when they were loose with the truth about the harmful effects of tobacco, they're doing the same again with vaping. But then, when has Big Tobacco ever been honest?

Who's really paying for Australian Big Tobacco?

In Australia, tobacco causes about 80% of all drug-related deaths.

The revenue from taxing tobacco is about $5–6 billion each year. But the total cost from tobacco to the Australian taxpayer is about $31 billion each year.

That gap of $25 billion each year is a monumental chunk of money. Wouldn't it be nice if Big Tobacco stepped up and paid the tab that rightfully belongs to them?

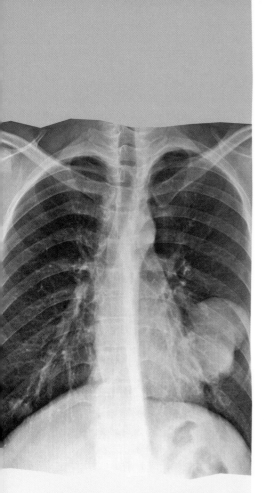

SMOKE AND MIRRORS –
TOBACCO DENIALISM

Cigarette packets today show graphic images of yellow-stained teeth, ulcers and lung cancers. They are as far as you can get from the glossy, glamorous cigarette advertising that was commonplace back in the 1900s. But getting to our current, more realistic representations of smoking has been a big fight – every single step of the way.

For most of the 20th Century, tobacco advertising was all about associating smoking with a seductive lifestyle, good health, outdoor activities, weight loss, and even endorsements from medical doctors. How on Earth did the tobacco companies ever get medical doctors to endorse cigarettes?

Today, we know that cigarettes can kill. In the year 2015, lung cancer, overwhelmingly caused by cigarette smoking, killed 1.7 million people worldwide.

If you knew the history, you would certainly think that the matter of whether smoking was bad for your health had been settled over half a century ago. Back in 1964, the US Surgeon-General released a landmark report stating quite resolutely that smoking definitely caused lung cancer. This finding was based on more than 7,000 published scientific papers. (As an aside, since 1964, the number of Americans killed by tobacco is 20 million. That's about fifteen times greater than the number of Americans who died in all of their wars, all added together.)

But Big Tobacco has never cared about the facts. On one hand, in 1999, Philip Morris, then the USA's biggest cigarette maker, responded by acknowledging that smoking caused lung cancer and other deadly diseases. But the giant British company Imperial Tobacco took the opposite approach. In a 2003 court case they filed documents stating, 'Cigarette smoking has not been scientifically established as a cause of lung cancer. The cause or causes of lung cancer are unknown.'

A SHORT HISTORY OF SMOKING

In 1604, King James I said that smoking was 'loathsome to the eye, hatefull to the nose, harmful to the braine [and] dangerous to the lungs …'

'Dangerous to the lungs'? How dangerous? A persistent cough is

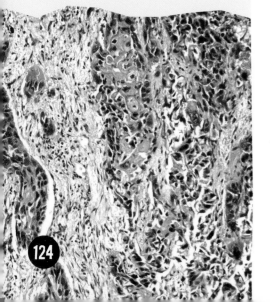

Health foundation?

The huge tobacco company Philip Morris International has set up a 'non-profit' foundation called The Foundation for a Smoke-Free World.

That's on a par with Big Fossil Fuel setting up a foundation to help the Great Barrier Reef.

one thing, but lung cancer is quite another. It wasn't until the 18th Century that the medical profession even recognised lung cancer as an official disease. In fact, by the year 1900, there were only about 140 cases of lung cancer in the entire published medical literature.

Around 1900, cigarette smoking was still relatively uncommon. It existed on the cultural periphery, kind of similar to having a tattoo last century. But within a few decades, smoking became a normal, mainstream part of the culture. It had become 'one of the most popular, successful, and widely used items of the early 20th Century'.

How did Big Tobacco make this colossal social and cultural change happen? The key was massive advertising that used celebrities, authority figures and influential public figures to promote and lend credibility to their products.

But then, something bad happened. To the puzzlement of the doctors at the time, in the 1920s, an epidemic of lung cancer had begun to appear – and the doctors didn't know why. There were a number of suspected causes. These included the recent global influenza pandemic of 1918–1919, industrial air pollution, massive amounts of tar laid on the new roads for the new motor car, exposure to poison gas in World War I, mouthwash, salted fish and – you guessed it – smoking.

PR SMOKE SCREEN

As the years rolled by, it became clearer that the lung cancer epidemic was getting much worse, and that cigarette smokers were suffering other health effects as well.

The tobacco companies ignored the health issues and instead responded with a very successful advertising strategy. The strategy had three major components. First, each company would link their own specific brands with health advantages, as compared to their competitors' brands. Second, they would get medical doctors to endorse their cigarettes. And third, they cast doubt on all scientific studies discussing the health effects of smoking.

So, back in the late 1920s in the USA, the Lucky Strike brand of cigarettes was the first to incorporate medical doctors in their ads, such as '20,679 Physicians say "LUCKIES are *less irritating*"'.

By the mid-1930s, Philip Morris, then a newcomer to the tobacco industry, was claiming that, after smoking their cigarettes, 'every case of irritation cleared completely, and definitely improved'. This helped Philip Morris transition into a major tobacco company.

Deadly time bomb

On average, cigarettes will kill one person for each million cigarettes smoked – but after a time delay of about 25 years from when they first started smoking. For example, the six trillion cigarettes that were smoked in 1990 killed about six million people in 2015, 25 years later. About one quarter of these people died from lung cancer, while the remainder died from other causes related to cigarette smoking.

To get a sense of scale, imagine your average 12-metre shipping container carrying ten million cigarettes. These cigarettes will kill about ten people – after about a quarter of a century.

By the 1930s, tobacco ads first began appearing in prestigious medical journals. Cigarette smoking was now widespread and in society, and the majority of medical doctors also smoked. Of course, the tobacco companies gave free cigarettes to doctors, especially at medical conferences.

But then the science began rolling in. By 1939, the first proper case-control study showed that people with lung cancer were far more likely to have smoked cigarettes than people without lung cancer. After World War II, the disconnect became more apparent. The tobacco ads showed *more* doctors apparently endorsing various brands.

But more and more medical studies appeared, showing the harmful effects of cigarette smoking. By 1950, five major statistical studies from the USA and UK confirmed that practically all cases of lung cancer occurred in smokers. This was big. In 1953, both *Life* and *Time* magazines devoted many pages to saying that the link between cigarettes and lung cancer was now proven 'beyond any doubt'. The *Journal of the American Medical Association* even stopped accepting cigarette advertising.

So, on 14 December of that year, the six biggest cigarette manufacturers in the USA met at the Plaza Hotel in Manhattan. They had to overcome their commercial rivalry to work out a combined master plan. This was a real crisis for the tobacco companies. Cigarette sales, and their share prices, were actually dropping. They had to do something different – and fast. So, they cancelled all Christmas holiday plans and engaged the USA's leading public relations company, Hill & Knowlton.

FAKING THE NEWS

The giant PR firm quickly realised that rejecting point-blank the torrent of scientific facts wouldn't work. So rather than simply *deny* the science, the key factor was to *control* the science. They messed around with the information that was getting to the public.

They claimed the real story was that there was controversy over the 'findings' about cigarettes being harmful to health. They just plain lied, and said there were still two sides to the 'hysteria linking smoking to cancer', so the issue (they claimed) was still in doubt.

They called for new research. Brilliant strategy. Just making this simple statement suggested both that they were open to new results, and that the existing research wasn't good enough. For this new approach to work, all the tobacco companies had to join together, to present a united front. And the PR firm created and funded a fake research organisation to sponsor medical research into tobacco.

In 1953, Big Tobacco set up the TIRC, the Tobacco Industry Research Committee. Shortly after, they released a full-page advertisement in more than 300 newspapers across the USA, known as the 'Frank Statement'. This ad claimed that it had not been proven that smoking had bad effects on health, and that the Tobacco Industry would do some proper research – but, overall, that the Tobacco Industry would look after the health of its customers.

The first director of the Tobacco Industry Research Committee was the noted geneticist Dr Clarence Cook Little. He held a very minority view that smoking was not linked to lung cancer. Little reckoned lung cancer was not caused by smoking, but by your unfortunate genetics, with maybe a little help from hormonal and emotional factors. Having such a prestigious academic as a figure head certainly helped add weight to Big Tobacco's statements.

The Tobacco Industry Research Committee was very successful at muddying the minds of the general public for a few more decades, but they didn't succeed as well with the medical profession.

Doctors began to quit smoking. In 1954, in Massachussetts, about 52% of physicians were smokers. Some 30% of all doctors were smoking a packet or more each day. Within five years, by 1959, only 39% of doctors were smokers, and only 18% of doctors smoked at least a pack per day. So doctors, who were most 'in the know', were quitting smoking in increasing numbers, and those who did smoke were smoking less.

But it took time even for the most educated (in regards to health) to fully accept the research about smoking. In 1960, only one third of all US doctors accepted that the case against cigarettes had been fully established. Eventually the science prevailed, and by 2006, only 4% of US physicians smoked cigarettes.

But the lessons of how to successfully sway public opinion turned out to be a goldmine for Big Tobacco – and others. The core message was 'Doubt is our product'.

Big Tobacco was the first to effectively use this tactic. The technique was later successfully used by various governments to justify invasions, by pharmaceutical companies, and many other organisations trying to put a spin on the truth.

And don't even get me started on climate change deniers ...

So it's still safe to say that the Merchants of Doubt are doing a roaring trade ...

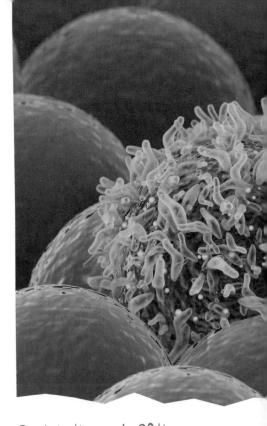

Back in the early 20th century, when the medical profession allowed itself to be associated with Big Tobacco, it hadn't yet been fully proven that cigarettes were bad for your health.

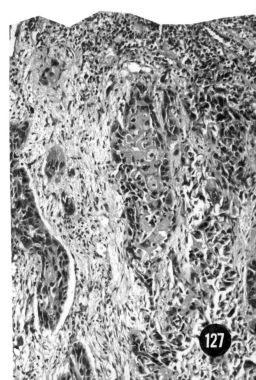

A2 VS A1 MILK

Gone are the days when you could buy warm milk in a bucket, fresh from the cow. Also gone are the days when there were just two milk choices – homogenised or non-homogenised.

Nowadays, in addition to 'skim milk', we have 'added calcium milk', 'light milk', 'powdered milk', 'UHT milk', and don't forget 'chocolate milk' – all of which have added time to how long it takes to cruise the dairy section at your local supermarket.

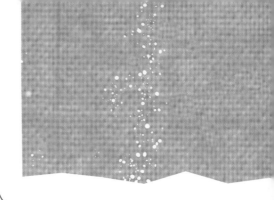

And then 'mylks' marched into town. These are plant-based, non-dairy milks that never saw the mammary gland of any animal. Soy milk was probably one of the first, followed by milks made from nuts, such as almond, cashew and macadamia.

It's all very confusing.

But to really ramp up the Bewilderment Factor another level, there's an expensive addition to the milk fridge called A2 milk (which at least comes from a cow). A2 milk is touted to be better than boring old A1 milk.

But, you guessed it, the science behind the supposed 'benefits' of A2 milk is a little murky.

Initial advertising for A2 milk claimed that 'regular' milk contained A1 proteins that would break down into a harmful 'peptide'. The A2 company alleged that this breakdown product of ordinary milk may cause diseases ranging from autism and schizophrenia to Type 1 diabetes and heart disease. This is amazing. How come nobody noticed this before?

But what even is A1 or A2 milk?

Let's start at the beginning.

'Regular' milk has a combination of both A1 and A2 proteins. The exact ratios of A1 and A2 depend on the breed of the cows, where in the world the cows live, and so on.

By the way, out of all the non-animal 'mylks', almond milk is the closest to plain water. Put another way, it's the one most lacking in nutrients. How did we manage to take almonds, which are rich in nutrition, and turn them into liquid emptiness?

DIFFERENT TYPES OF 'COW'

Bos taurus (taurines) are what we know as the common cow. They produce most of the milk consumed in the Western world. But there's also *Bos indicus* (zebu), and *Bos grunniens* (yak). (So that makes the most important cow *Boss Bos!* 😉)

MILK AND HUMAN HISTORY 101

The story of humans drinking milk began thousands of years ago, thanks to a mutation to the human DNA.

Milk contains the carbohydrate lactose. It's made from two smaller sugars that are stuck together. Unfortunately, this lactose carbohydrate is too big to cross the gut wall. So, for most adults on Earth, lactose stays inside the gut – and does not get absorbed into the bloodstream. If you can't absorb lactose from your gut, it has to pass out in your poo. But on its way out to the toilet bowl,

lactose can cause various symptoms, such as gas, bloating, and even diarrhoea. (This is due to 'osmosis', which I discuss in **DO FISH DRINK WATER?** on page 148.)

But babies drink milk. How come?

Because most babies naturally make an enzyme called 'lactase'. This chemical splits lactose into two smaller carbohydrates (or sugars) that can cross the gut wall in the small intestine, and later be broken down to give energy.

This is great. Both babies and mothers can benefit from breast-feeding.

Breast milk gives babies everything they need, until about six months of age. (Up to six months, breast milk can provide all the nutrition a baby needs, including iron.)

For the mother, the first stage in the process of breast-feeding is that the brain releases the hormone oxytocin. Oxytocin (in the breast) starts the mechanical flow of the milk from the breast ducts. But oxytocin (in the uterus) also sets off gentle contractions in the uterus, which improves the uterus's muscle tone in the early weeks after birth. Breast-feeding is also free of charge, very convenient, helps some women lose weight – and builds up the bond between the baby and mother. An extra benefit is that each 12 months of breast-feeding reduces the risk of later breast cancer by about 4–5%.

So, if babies can make lactase (which splits lactose), why can't adults? What changes?

Well, in Ye Olde Days, babies lost the ability to make lactase soon after they were weaned. After all, why spend energy making a chemical you don't need?

But then a lucky mutation happened. It happened in Hungary around 7,000 years ago, and in Africa about 4,000 years ago.

The mutation meant that people began to make the enzyme 'lactase' – not just when they were infants, but also when they were adults. So now they could drink milk throughout their whole lives.

Back 7,000 years ago, this mutation gave a particular advantage to those who had both the mutation and some cows. In one year, a cow could give them, via milk, as much nutrition as if they had killed the cow and eaten it. But the advantage was that, at the end of one year, they still had the cow – and it didn't compete with humans for food. (Cows eat grass, which humans don't.)

This mutation gradually spread across the world, and now about one-third of the population has it. It's common in Europe and Africa, and uncommon in Asia – and, in between, you get a mix where some people have it and some don't.

Adults who don't make the lactase enzyme are 'lactose intolerant'. But many can still manage a small amount of milk. The majority of people in the world can drink a splash of milk in a cuppa without any trouble. On average, they can drink up to about 12 g of lactose (the amount in a cup of milk), but they can't drink a whole half-litre milkshake, or else their guts go 'messy' (i.e. they experience wind, diarrhoea, etc).

A2 TO A1

It is thought that back some 5,000–10,000 years ago, practically all milk was A2.

But as we bred the cows to give us more volume of a better-quality milk, as well as a greater carcass weight, our 'selective breeding' accidentally switched the milk from mostly A2 to mostly A1. So today, most cows give A1 milk; these include Friesian, Ayrshire, British Shorthorn and Holstein cows.

Milk that is mostly A2 comes from Southern French breeds of cow (Limousin and Charolais), from Channel Island cows (Jersey and Guernsey), and from the original zebu cattle of Africa.

And finally, buffalo milk is generally high in A1.

MILK NUTRITION 101

Cows' milk is about 88% water. But it does carry the three main chemical groups that are essential in our diets – carbohydrates (5%), fats (4%) and proteins (3%). (This specialised mix of lovely nutrients makes milk an ideal post-gym drink for lactose-tolerant gym bunnies.)

In A2 milk, one of its proteins is (very) slightly different from the corresponding protein in A1 milk. So we need to look at milk proteins.

Correlation is not causation

How on Earth did people begin to think that milk was bad for you? (I'm talking about the one-third of the population of the world who can digest milk.) Here's a possible explanation.

Back in 1992, R.B. Elliott (Department of Paediatrics, School of Medicine, Auckland, New Zealand) was interested in how the environment of Polynesian children affected their risk of type 1 diabetes. The risk was very low if they lived in their homeland of Western Samoa, but much higher if they lived in Auckland, New Zealand. There was a strong correlation between the rates of type 1 diabetes and consumption of cows' milk.

As far as he was concerned, there was only one possible answer – cows' milk. However, he was wrong, as there were many other confounding factors involved.

Several follow-up studies, including one in 2018, showed no link between childhood diabetes and milk.

PROTEINS 101

A protein is just a string of amino acids joined together.

Next question: just what is an amino acid?

An amino acid is a small molecule containing atoms of carbon, hydrogen, oxygen and nitrogen.

At one end, there's a group of three atoms called an 'amine' – NH_2 – that's one nitrogen atom and two hydrogen atoms. At the other end is a group of four atoms called a 'carboxyl group' – COOH, or one carbon atom, two oxygen atoms and one hydrogen atom.

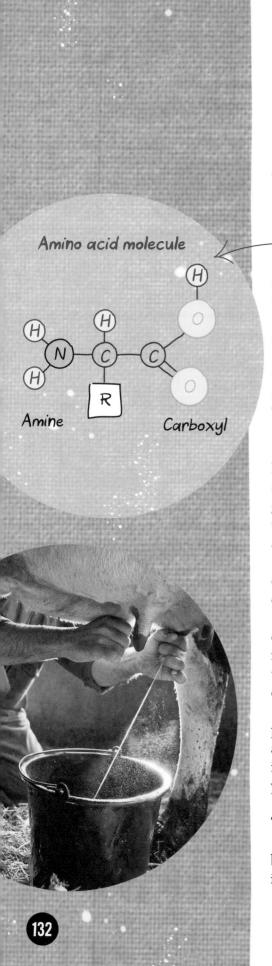

Amino acid molecule

Amine

Carboxyl

And in between, there's a side chain (or residue) – a bunch of atoms that are unique to each amino acid.

Amino acids are essential in the human body. They give energy, they can carry neurotransmitters, and they're the building blocks for other useful chemicals (e.g. serotonin and haem, as in haemoglobin). The 20 common amino acids were discovered between 1806 (asparagine) and 1935 (threonine).

Biochemists know of some 500 amino acids. Human DNA can handle only 20 different amino acids (but joined together in many complex combinations). Typical human amino acids include arginine, glycine and proline. Because we can't make enough of about eight of these 20 amino acids, we have to eat them. (These are called 'essential' amino acids, and include leucine, isoleucine and valine.)

If you stick any two amino acids together, you call the resulting chemical a di-peptide. If you stick three amino acids together, you call it a tri-peptide.

As you stick more and more amino acids together, it's still called a peptide until you get up to about 50 amino acids. And once it's longer than about 50 amino acids, it's called a protein. (Sorry, this is a fuzzy area – some biochemists pick '20 amino acids' as the changeover point between a peptide and a protein.)

Peptides can have all kinds of effects on the human body. You've probably heard of various athletes taking peptides for bodybuilding and bulking up. Peptides can affect your blood pressure and your immune system, fight bacteria, bind minerals and even influence how your blood clots. You get the idea: peptides can be quite powerful.

Proteins have essential structural and functional roles in our bodies. They are also a source of energy, nitrogen and essential amino acids.

It turns out that all of our food groups have proteins that are broken down into various bioactive peptides. These food groups include cereals, vegetables, meat, and, yes, dairy.

This gets us closer to A2 milk.

CURDS AND WHEY & MISS MUFFET

Even Little Miss Muffet (that poor little arachnophobe) realised that dairy products are rich in nutrients.

We don't know who originally wrote it, but this nursery rhyme first appeared in print in 1805. It runs:

Little Miss Muffet,
Sat on a tuffet,
Eating her curds and whey,
Along came a spider,
Who sat down beside her,
And frightened Miss Muffet away.

In milk, about 80% of the proteins are insoluble, but 20% are soluble (they'll dissolve in water).

'Curds' are loaded with the insoluble proteins in milk. They are the semi-solid lumpy bits that you get by curdling milk. You can curdle milk by adding rennet, or something acid (like vinegar or lemon juice). Curds are rich in casein proteins. Making curds is one of the early steps towards making cheese.

'Whey' is the liquid that contains the soluble proteins. Don't underestimate whey. This valuable liquid contains many important chemicals such as alpha-lactalbumin, beta-lactoglobin, albumin, and lactoperoxidase. They have important nutritional and biological properties, including antimicrobial and antiviral actions, immune system stimulation, anticarcinogenic activity, and much more.

MILK PROTEINS – 7 LEVELS TO DIVE DOWN

A litre of whole milk contains about 32 g of protein. (But milk that is reduced in fat has about 36 g of protein.)

To get to the difference between A2, A1 and A3 proteins, we have to dive down about seven (7!) levels.

Level 1	Let's start with the (roughly) 32 g of protein in whole milk.
Level 2	Let's now look at just one-third of all the proteins in milk – the casein family of proteins. (As explained in the Little Miss Muffet box, caseins are the 'curds', the semi-solid lumpy bits that you get by curdling milk.)
Level 3	There are four different varieties of casein. They are as1, as2, beta-casein, and kappa-casein. Now consider just the beta-caseins.
Level 4	The beta-caseins exist in three different varieties called A, B, and C. (A and B are the most common.) Let's zero in on the A variety.
Level 5	The A variety can exist in three different sub-varieties – A1, A2 and A3. They each contain 209 amino acids. Let's dive right in and look at just A1 and A2.
Level 6	A1 and A2 proteins are almost identical – except in just one single amino acid.

Remember those 209 amino acids in Level 5, in Type A beta-casein protein? They are conveniently numbered by their position – from 1 at one end, to 209 at the other end.

The only difference between A1 and A2 beta-casein is the amino acid at position 67. Here, A2 has a proline amino acid, while A1 has a histidine amino acid. (This was first discovered way back in 1961.) That is the total (and only) difference between them.

Supposedly evil

$$... -Tyr^{60}-Pro^{61}-Phe^{62}-Pro^{63}-Gly^{64}-Pro^{65} -Ile^{66}-(His^{67})... \; \beta-casein \; A^1$$

$$... -Tyr^{60}-Pro^{61}-Phe^{62}-Pro^{63}-Gly^{64}-Pro^{65} -Ile^{66}-(Pro^{67})... \; \beta-casein \; A^2$$

Supposedly not evil

Level 7	Once in the human gut, various enzymes break down the A1 protein, at position 67, into a 'supposedly evil' peptide. However, the presence of a different amino acid, proline, at position 67 in A2 milk stops this breakdown to the supposedly evil peptide. The incorrect claim (that this peptide is 'bad for your health') is the single 'rock' upon which the whole A2 financial empire is built.

RATIOS OF A PROTEINS

All cows generate a mixture of mostly A1 and A2 proteins, as well as a smattering of B and other proteins. The important thing to realise is that, around the world, the ratios of these A1 and A2 casein

proteins in cows' milk vary enormously. No cow delivers 100% A2 milk (or 100% A1, or 100% B).

First, the A1 cows. The DNA of Angus cows is about 95% A1 (and 5% B). Brahman cows' DNA is about 99% A1 (and 1% B). Ayrshire cows in the USA are about 72% A1 (and 28% A2), but in the UK and Canada, they are 60% A1 (and 40% A2). In New Zealand, this drops to 43% A1 and 53% A2.

Now for the predominantly A2 cows.

The A2 DNA level reaches 97% in Guernsey cows in the USA (with the remainder being 1% A1 and 2% B).

Holstein cows in Germany are 96% A2 (and 4% B). But Holstein cows in Ireland are just 25% A2 (and 72% A1, and 3% B).

Now you can see that it's not as simple as the A2 marketing would have you believe.

SO WHAT?

Finally, we are back at the supposedly evil peptide.

This A1 peptide (made from seven amino acids stuck together) is wrongly claimed to cause everything from autism to schizophrenia, and type 1 diabetes to heart disease.

Yes, it is true that some peptides can have powerful effects on us humans.

And yes, with A1 milk, the gut enzymes manufacture a bioactive seven-amino-acid peptide called beta-casomorphin, or BCM-7. As you can guess from the 'morphin' part of the name, this chemical is related to the opiate family. But it's a very weak opioid.

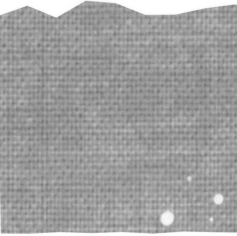

$-\text{Tyr}^1-\text{Pro}^2-\text{Phe}^3-\text{Pro}^4-\text{Gly}^5-\text{Pro}^6-\text{Ile}^7$
$\dots \beta\text{-casomorphin-7}$

This 'evil' bioactive peptide is supposed to cause all the harmful effects of drinking regular A1 milk.

This peptide, BCM-7, has been found to have a wide range of effects in animals, but only if injected directly into the blood. When injected directly into rats, BCM-7 has been found to have pain-killing activity, and will also accelerate learning of a food-procuring habit. (The obvious lesson is not to inject milk into your bloodstream. 😉)

Moreover, it appears that only peptides up to about three amino acids long can cross the gut wall. So a peptide that is seven amino acids long won't cross the wall of the gut. Indeed, although scientists have looked for the presence of BCM-7 in human blood after drinking cows' milk, it has never been found. BCM-7 has been shown to have minor effects on gut movements and inflammation in animals – but again, this has not been shown in humans.

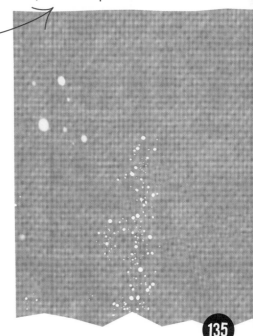

These weak and inconclusive findings about BCM-7 are the (incorrect) basis of the claims that A1 is bad for you.

SHOW ME THE MONEY!

The A2 Corporation was formed in 2000 to commercialise A2 milk. They had a patent on a genetic method for identifying cattle that would produce A2 milk.

(The weird thing is that their test did not check the final product, the actual milk, for how much A2 protein was present. If they're selling milk that is supposedly loaded with A2 protein, why do they test the cow, but not the milk?)

In 2002, the A2 Corporation launched a High Court lawsuit against a major New Zealand milk cooperative (Fonterra Cooperative Group), accusing it of covering up the possible harmful effects of A1 milk. (At that time, Fonterra generated 20% of New Zealand's total offshore receipts. It's also New Zealand's largest company, and accounts for about 30% of the world's dairy exports.) In 2003, the A2 website (incorrectly) claimed that 'Beta casein A1 may be a primary risk factor for heart disease in adult men, and may also be involved in the progression of insulin-dependent diabetes in children'.

The A2 Corporation actually wanted health warnings placed on all containers of A1 milk, warning the innocent consumer of 'regular milk' of these risks. So they petitioned the Food Standards Australia New Zealand regulatory authority to have these health warnings added.

However, none of the research that backed up these very scary claims was solid. To summarise, it was based on small studies, which were badly done, usually not relevant to humans, carried out for too short a period of time, and usually not published in peer-reviewed journals. They were often studies comparing different countries, or comparing a few environmental factors with some non-communicable diseases. Indeed, a lot of self-reporting was involved.

ALL A2, NO A1? NO

The A2 Corporation (which later changed its name to The a2 Milk Company Limited) originally claimed that its A2 milk contained absolutely no A1 protein.

However, in 2003, the New Zealand Commerce Commission found some A1 protein in what was supposedly 100% A2 protein. As a result, it stopped the A2 Corporation from claiming that its A2 milk was free of A1 protein.

SHOW ME THE SCIENCE!

Back in 2005, Professor A.S. Truswell (Human Nutrition Unit, University of Sydney) carried out a comprehensive investigation. He wrote that the overwhelming scientific evidence was that there was 'no convincing or probable evidence that the A1 beta-casein protein in cows' milk is a factor causing' type 1 diabetes or heart disease. His report went on to say that the evidence for A1 beta-casein proteins in milk causing autism and schizophrenia 'is more speculative, and the evidence is more unsubstantial than that for DM-1 and for CHD [Type 1 Diabetes Mellitus and Chronic Heart Disease]'. To summarise, 'there is no convincing or even probable evidence that the A1 beta-casein of cow milk has any adverse effect in humans'.

And there you have it.

In 2005 there was zero proof that A1 milk was bad for you, and zero proof that A2 milk was better than A1. (In 2009, the European Food Safety Authority published a more comprehensive 107-page multi-author scientific report on the health effects of A1 and A2 milks – and reached the same conclusion as Truswell. It's still the same in 2019 ...)

As you would expect, today the claims on the 'superiority' of A2 milk have been much softened. They are much more vague and typically refer to gut discomfort (e.g. bloating and flatulence) after drinking milk. But, as before, the studies showing the bad effects of A1 versus A2 milk are small, and need improvements. Many of the studies were carried out in people who were known to have lactose intolerance, with typical symptoms of bloating and flatulence.

Milk is a good all-round food, and is an especially rich source of calcium that is very bio-available. This means that your body can absorb lots of the calcium in milk.

A2 milk is more expensive, partly because of the cost of carrying out genetic testing for five years.

So, to summarise, the health benefits of A2 milk seem to be close to zero, while the cost is much higher. Following on from their slogan of 'Feel the Difference', it's likely that the only difference you may feel may be in your wallet.

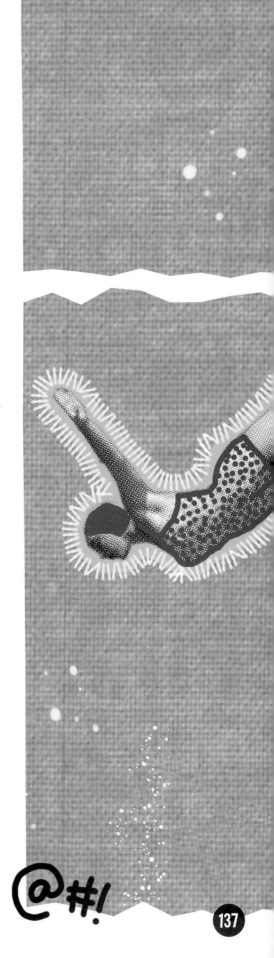

KISS THE SUN

Without the Sun shining continuously, things on Earth would be rather bleak. Our planet would cool down by a few hundred degrees. Almost all living things would very quickly die out. Without the Sun's gravitational pull, our Solar System's planets would scatter into space, or possibly smash into each other.

So it's fair to say our Sun is essential to life as we know it.

And yet, despite humanity spending much of recorded history trying to understand our nearest star, the Sun, there is still so much we don't know about it. This is why, on 12 August 2018, NASA launched a spacecraft to take a closer look. The Parker Solar Probe has been designed to actually skim through the Sun's superhot outer atmosphere (the corona) and make observations along the way. It will 'kiss the Sun' (well, almost).

SUN 101

I love the magnificent engineering marvel that is the Parker Solar Probe. But first, a little background on our Sun.

The Sun is enormous compared with the Earth – about 109 times bigger in diameter, and about 330,000 times heavier. The Sun carries about 99.86% of the mass that exists in our Solar System. The temperature at the central core of the Sun is a titanic 15.7 million °C. The Sun generates its enormous heat output by burning, each second, about 600 million tonnes of hydrogen, and turning that into 596 million tonnes of helium.

But our mathematically inclined readers will have noticed a missing 4 million tonnes each second in that last computation.

Well, about 3 of those 4 million tonnes get mostly turned into energy, via Albert Einstein's famous $E = mc^2$ equation, where 'E' is

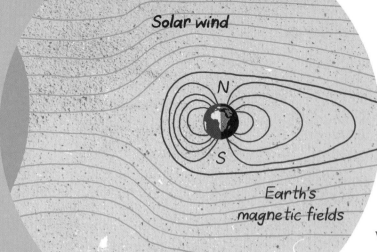

Solar wind

N
S

Earth's magnetic fields

the energy, 'm' is the mass, and 'c' is the speed of light.

That leaves about 1 million tonnes unaccounted for. These are mostly charged particles, with mass. They are ejected from the Sun, in the form of what we call the solar wind.

The solar wind is coming from the Sun on the left, is squashed a little on the upwind side of the Earth – the ball that is half white (for daytime), half black (for night-time) and labelled N & S (for North and South) – and stretched out into a long tail on the downwind side (away from the Sun).

SOLAR WIND 101

The solar wind is a fairly constant torrent of charged particles, leaving the Sun and heading outwards in all directions, and blasting out past Pluto. They are mostly electrons, protons and alpha particles (helium nuclei).

The solar wind is moderately constant in temperature, speed and density – but it does vary a little. This variation relates to time, and location on the Sun.

The solar wind comes in two distinct streams – slow and fast. The slow stream tends to originate from the Sun's equator and is doing 300–500 km/sec by the time it reaches Earth.

The faster stream gets to us at around 750 km/sec, and seems to come more often from coronal holes near the Sun's poles. Coronal holes are regions on the surface of the Sun that are dark and comparatively cool.

We don't know the cause of the slow solar wind, but we have a better understanding of the fast kind.

The solar wind rams into the Earth's magnetic field and squashes it on the Sun-facing side. This leaves a long magnetic tail reaching out from the night side of the Earth.

But while Earth has a protective magnetic field, Mars does not. It seems that the solar wind has stripped away its atmosphere over billions of years. Today, the atmosphere of Mars is just 0.6% of Earth's.

It's also thought that the solar wind blowing out from the Sun forces comets to have their tail pointing away from the Sun.

So that's part of what we know. Let's look at what we don't know about our toasty neighbourhood star.

This shows a comet's tail stretched out and facing away from the Sun

Mars
Venus
Sun
Mercury
Earth

IT'S A MYSTERY

Mystery number 1: part of the outer atmosphere of the Sun, called the corona, is much hotter than the surface. That sounds the wrong way around – it should be cooler.

The temperature in the corona is a few million degrees, versus some 5,500°C for the surface. That's about 300–450 times hotter! If all the power is generated in the centre of the Sun, and if everything gradually cools down towards the surface, how can the temperature suddenly then rise again once we get beyond the surface? We don't know. That's one mystery.

Mystery number 2: how does the energy that heats the corona manage to accelerate the high-speed particles of the solar wind? Some of them can reach astonishingly fast speeds – up to half the speed of light! We think magnetic fields are involved, but how? We still don't know the pathway by which it all happens.

These questions are not just idle curiosity. Their answers affect us down here on the ground.

On 9 March 1989, the Sun had a hissy fit and threw out a huge superhot ball of solar wind (a coronal mass ejection). It smashed into the Earth's atmosphere, and set off a severe geomagnetic storm that overwhelmed Earth's 'regular' magnetic field. This event crashed the electric power grid around Quebec, in Canada, for nine hours.

These mysteries, and the Sun's tempestuous history, are why NASA thought it'd be a good idea to send in the Parker Solar Probe.

This spacecraft is an incredibly impressive engineering feat. A huge amount of time, money, brainpower and raw power has gone into figuring out how to get this car-sized robotic probe to kiss the Sun without it melting and/or being sucked in and swallowed up by the Sun's gravitational pull.

GRAVITATIONAL SLINGSHOT

Even something as seemingly 'simple' as getting a spacecraft close to the Sun is not easy. You might think that all we need to do is get it into orbit, point it in the right direction and hit the big red Launch button.

But, no. Surprisingly, it takes 55 times as much energy to get to the Sun (and then not smash into it), as it does to get to Mars!

This sounds very weird. First, let's think about going to Mars (and then we'll think about going to the Sun). In each case, we have to get away from the Earth's surface. To do this, we need our rocket to reach a velocity of 11 km/sec.

Then, to get to Mars, we fire another rocket to just add another 2 km/sec. The rocket scientists call this 'delta v' – the change in velocity. And nine months later, we arrive at Mars.

Second, let's think about going to the Sun. This is much harder – instead of adding a delta v of 2 km/sec, we have to 'scrub off' (lose velocity) lots of delta v.

Just as with going to Mars, to leave the Earth's surface our rocket has to reach 11 km/sec. Even though we have left the atmosphere, we are now in orbit around the Sun, still travelling at 30 km/sec in company with Earth.

But think about this – even though the Sun is big and has a huge gravitational field, the Earth does *not* fall into the Sun. This is because the Earth is orbiting around the Sun with a 'sideways speed' of about 30 km/sec. To spiral in a bit closer to the Sun, we have to scrub off some of that 30 km/sec of speed.

However, to get really close to the Sun (the Parker Probe is aiming for 6.2 million km from the Sun's surface), we have to slow the craft down by a huge amount. The mathematicians tell us we have to somehow scrub off 23.7 km/sec of velocity. The only way we can remove that velocity is with a rocket. (It's kind of weird to think you need to fire a rocket to slow down, but here's how it works. If the rocket is pointing in the direction of travel, when you fire the rocket engine, the rocket will speed up. But if you turn the body of the rocket around 180°, and then fire the engine, the rocket will slow down, while still travelling in the same direction.)

Sadly, even our most powerful rockets, with a very light payload, can't give that payload a delta v of 23.7 km/sec.

The rockets used in deep space missions normally can achieve a delta v of only 7–10 km/sec. Even the fastest mission ever launched – the very light New Horizons space probe to Pluto, launched by our most powerful rocket – left Earth at only 16.26 km/sec.

The Parker Solar Probe is not a heavy spacecraft – only about 685 kg. We launched

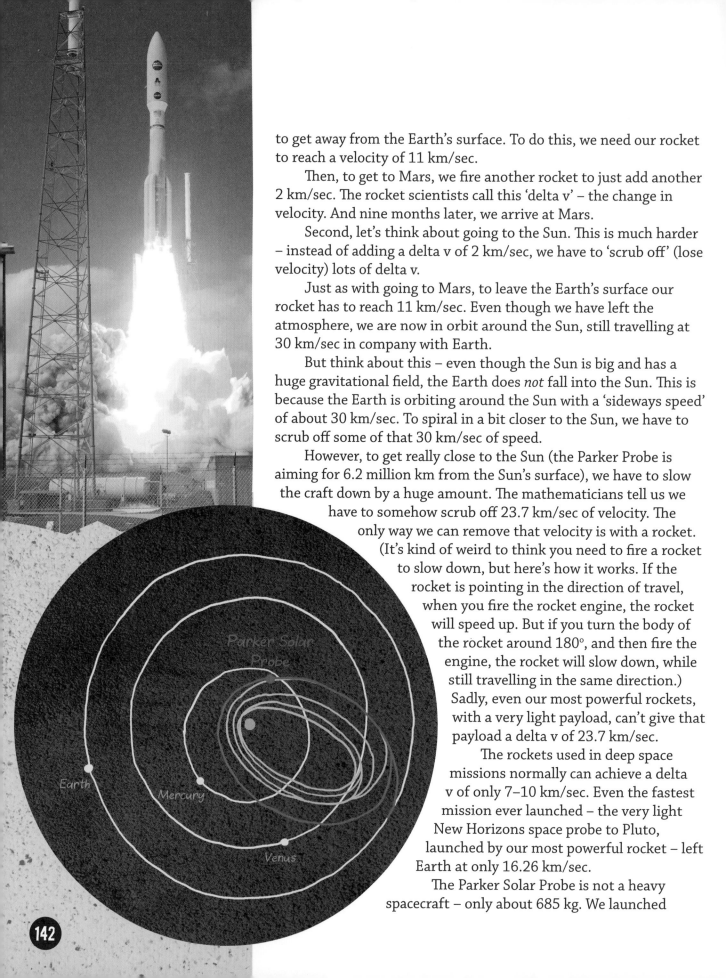

Parker Solar Probe

Earth

Mercury

Venus

it with one of our most powerful rockets (the American Delta IV Heavy) – normally used for heavy payloads – to try to shave off as much as possible of its excess velocity. But even using our most powerful rocket on a light spacecraft wasn't enough to close that 23.7 km/sec velocity gap.

So, how will the Parker Solar Probe scrub off its excess velocity? Answer – with a series of gravitational assist (or gravitational slingshot) flybys of Venus. In the past, we have used this manoeuvre to speed up spacecraft, to visit the outer Solar System. It can also be used in reverse to slow a spacecraft down, to visit the inner Solar System.

In each flyby, the Parker will transfer some of its energy to Venus. So the probe will slow down, but the planet Venus will speed up in its orbit around the Sun (but only by a truly microscopic amount).

In its first gravitational slingshot manoeuvre, on 3 October 2018, the Parker scrubbed off 3.1 km/sec of its own orbital speed, as it passed within 2,500 km of the surface of Venus. The second flyby of Venus (26 December 2019) is expected to scrub off another 2.9 km/sec. The Parker will do this cosmic dance with Venus a total of seven times over seven years. It will zip past the Sun, at closer and closer distances, some 24 times.

Timeline of Parker Solar Probe

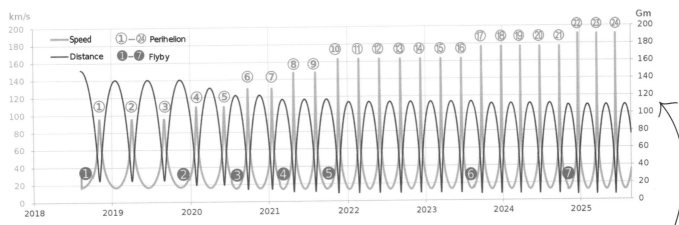

Gradually, the spacecraft will also change its orbit into a more elliptical (or football-shaped) orbit. At its closest to the Sun, in 2024, it will zip through the star's superhot atmosphere, just 6.2 million km from the surface – much, much closer than Mercury, the closest planet to the Sun. From there, it'll be able to send us some very precious data. There is no other way we can gather that data.

But hang on, how will it survive those hellish conditions?

On the right side, the 'gm' stands for gigametres, which is sciency talk for 'million kilometres'. It's the distance of the spacecraft from the Sun – the centre of the Sun.

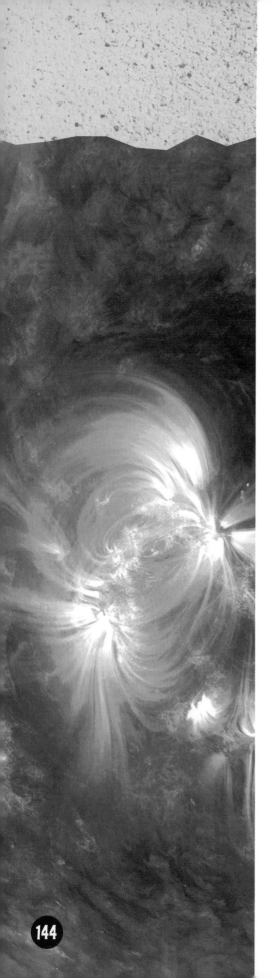

SURVIVING HELL

At a distance of just 6.2 million km, the power from the Sun's fiery surface is massively concentrated.

The Sun emits about 50% of its power as heat (or infrared light), about 40% as visible light (the colours of the rainbow from red to violet) and about 10% as ultraviolet light (that fraction of sunlight that makes both Vitamin D and skin cancers). If you measure this power near the Earth (about 150 million km from the Sun), each square metre of space is receiving about 1,400 watts of Sun-power.

But when the Parker Solar Probe is at its closest to the Sun's surface (6.2 million km), each square metre of the spacecraft that is facing square-on to the Sun will receive about 475 times more power – roughly 650,000 watts.

This is a truly hostile environment. That much power would kill an unprotected satellite within a minute or two.

To survive, the Parker will lurk in the shadow of its own sun shield. It will point its head directly at the Sun, hiding its rear end behind the shadow cast by a sophisticated heat protection shield. This shield is hexagonal, about 2.3 m across, and 11.4 cm thick (a bit fatter than your clenched fist). The shield is made from a reinforced carbon composite, and has a white reflective alumina ceramic surface on the side closest to the Sun. On the Sun side, the shield can withstand temperatures of 1,370°C. The protective barrier will cast shade over the probe's instruments, keeping them at a comfy 30°C.

But this shielding trick is further complicated by the Parker's flightpath. The craft will be in an elliptical orbit. When it's at its closest approach to the Sun, it will become the fastest ever human-made object, reaching almost 200 km/second, or about 700,000 km/hour.

Now, because it's travelling so fast, the good news is that it won't spend too long baking in the fierce heat (at its closest approach). But it will have to keep adjusting the orientation of the heat shield, to protect the probe with its shade. It will have to do this with its own internal artificial intelligence, because any commands from Earth could take eight minutes or more to get to Parker. In fact, Parker is the 'smartest' spacecraft we have ever sent (as of 2019).

FOUR INSTRUMENT SUITES

PARKER SOLAR PROBE

A MISSION TO TOUCH THE SUN

The Parker Solar Probe has some very important jobs to do on its journey to visit the Sun, so it can't spend all its time hiding behind its heat protection shield. Its jobs are to measure the Sun's electric and magnetic fields, to measure particles from the Sun and the Solar Wind, and to capture images of the environment around the probe itself.

The Parker carries four major sets of instruments. The first is probably fairly obvious: a pair of telescopes to look at the wispy, super-hot, atmosphere of the Sun. It's called WISPR (Wide-field Imager for Solar Probe). One telescope has a 40° field of view, while the other has a 58° field of view.

Then there's FIELDS – three magnetometers and five voltage sensors, to investigate the Sun's electromagnetic fields. They're mounted on antennae made from Niobium C-103, a special high-temperature alloy that can withstand the intense heat of the Sun. (Niobium C-103 is used in rocket engines, because it can maintain its high strength at temperatures of up to 1,482°C.)

ISOIS (Integrated Science Investigation of the Sun) and SWEAP (Solar Wind Electrons Alphas and Protons) will measure various aspects of the Sun's solar wind. ISOIS will investigate high-energy particles that are linked to solar activity, such as solar flares and coronal mass ejections. SWEAP will use its two Solar Probe Analysers (to measure electrons and ions in the Solar Wind), and its Solar Probe Cup (to catch the solar wind directly as it streams off the surface of the Sun).

Some of these sensors won't be hidden behind the solar shield – they've been hardened to survive direct exposure to the insanely hot conditions. (That's their job – to go in harm's way.)

For example, the Solar Probe Cup (also made of Niobium C-103 and sapphire) will catch the solar wind as it streams off the Sun. This crafty cup should give us our first close-up measurements of the solar wind.

Until now, most of our measurements of the solar wind have been near Earth, some 150 million km from the Sun. This distance has given the various elements of the solar wind enough time to meld into each other, mixing together and cooling down.

So, the solar wind at Earth is very likely to be different from the solar wind near the Sun. That's why we're going to the Sun.

SOLAR POWER (LIKE ON YOUR ROOF)

To power its fancy instruments, the Parker Solar Probe has two very different sets of photovoltaic solar panels.

The 'regular' set extend out from the body of the craft. But they will survive only while the probe is more than about 35 million km from the Sun. That's roughly two-thirds of the distance between the planet Mercury and the Sun. That's very close to the Sun, but the regular solar panels will still work just fine. They would fail if they were closer to the Sun.

But once the Parker gets even closer, those regular solar panels are gradually pulled back into the shade. A smaller, second set of solar panels are left exposed, to turn the fierce sunlight into electricity. To survive, these special solar panels are water cooled, to keep them below 150°C.

EUGENE PARKER

This spacecraft is named after the US physicist Eugene Parker. He was the first to theorise the existence of the solar wind, and to come up with a decent explanation, back in the 1950s. Interestingly, this is the first time that NASA has named a spacecraft after a living person. And as a further hat-tip to Eugene, the probe also carries a memory card, with photos of him, and one of his most important scientific papers on solar physics.

This memory card also contains the names of some 1.1 million people. These are members of the public who wanted a little part of themselves to accompany the probe on the very first mission to touch a star – or kiss the Sun. And my name is on that list!

But, I have to say, I'm glad it's just my *name* flying with the Parker. That intrepid little probe has a long, lonely and dangerous seven-year journey ahead of it. At least it won't have to worry about getting too cold.

The Fisher Space Pen is famous for working in the micro-gravity of space, and for its star appearance in a *Seinfeld* episode. But maybe the Fisher Space Pen has had its time in the Sun, and it should move over to make room for the Parker Solar Marker?

DO FISH DRINK WATER?

DETOUR

Human wee per day

We humans generate about 1 ml of urine per hour for each kilogram of body weight. If you weigh 75 kg, each hour you will generate about 75 ml of urine.

So, we humans wee out about 2–3% of our body weight each day – on average.

In terms of body weight, we wee a little less than saltwater fish (4%), but much less than freshwater fish (20%).

Young kids are always curious. They ask about everything. Looking upwards, they'll say, 'Why is the sky blue?' And then, looking at a fish tank, the question is, 'Do fish drink water?'

Yes, all fish 'drink' water. They have to. That's how they get their oxygen. They suck water into their mouth, and then squirt it out through their gills (where the oxygen in the water is removed, and then sent into their bloodstream).

But what about gulping in the water – and then swallowing it? Only *S*altwater fish actually *S*wallow water.

Easy to remember:
S for *S*alt water
S for *S*wallow

WEIRD

Then it gets really weird. The saltwater fish that swallow water like crazy make hardly any wee at all – only 4% of their body weight each day.

So, what happens to the rest of the water they swallow? It goes *out* through their gills and skin, and into the salty ocean.

And to keep some balance in our strange Universe, freshwater fish do the opposite. They don't drink much water – but they do wee heaps, about 20% of their body weight each day.

But where does the water that makes up their urine come from? You guessed it – from the fresh water around them, and straight through their gills and skin *into* their body.

(Just as an aside, it's much harder for water to pass through human skin. But enough water can get into your sweat glands to make your fingers wrinkle after a long bath.)

So that's *what* is happening. But to understand *why* this is so, there are two slabs of background information that you need to upload into your brain.

First: 'We all need oxygen to live', and second: 'Osmosis'.

NEEDING OXYGEN TO LIVE

Let's start with oxygen.

We humans have it sweet. It's very easy for us to get our oxygen because we are surrounded by it – it's in the air all around us.

All we have to do is take a breath – and air simply rushes into our lungs. Oxygen makes up one-fifth of the air we breathe. Within seconds, deep inside our lungs, oxygen from this air flows across some 70 m² of a very thin membrane – and then straight into the bloodstream. (The distance between the air in our lungs and the bloodstream is about 1 micrometre for us humans, so the oxygen molecules go across very quickly.) Bingo! Problem solved, and the necessary oxygen has been shifted from the air into our blood.

Fish also need oxygen, but getting it is a lot tougher for them. They cop a double whammy.

First, they live in water, which is about 800 times heavier than air. Moving heavy water around is a lot of work.

Second, there's about 30 times less oxygen in water than there is in the air.

The combination of these two factors means that fish have

Are the Amazon forests really the lungs of the Earth?

The answer is 'no', but the 5.5 million square kilometres of the Amazon do contribute a reasonable amount of the oxygen we breathe – as well as rain, massive biodiversity and more. So the Amazon does do a great job. But the majority of the oxygen we breathe comes from floating plants in the sea called 'phytoplankton' – 'phyto' meaning 'plant', and 'plankton' from the Greek word for 'wanderer'.

These phytoplankton in the oceans contribute about 50–85% of the oxygen we breathe, so land plants give us 15–50% of our oxygen. You might think that 50–85% is a pretty large variation – and you'd be correct. Part of the uncertainty is that while we can work out how much oxygen a single phytoplankton gives out, we don't know how many phytoplankton are in the oceans. They can float in just a thin film near the surface, or can occupy a layer 100 m thick – and to complicate things a bit more, they come and go with the seasons. So they can sometimes erupt in vast blooms in springtime.

	Water	Air
Weight/cubic metre	1,000 kg	1.2 kg
Oxygen/litre	7–10 mg/L	286 mg/L

to work a lot harder than we do to get essential oxygen into their bloodstream. Amazingly, the energy cost of getting oxygen out of the water varies from 10–70% of a fish's total metabolic energy needs. (The 70% happens when they are 'exercising' very vigorously – like getting the heck away from Jaws.)

Maybe the difficulty in getting oxygen was part of the reason animal life began to leave the oceans about 400 million years ago, to migrate onto the land?

Yet, even with the odds against them, fish have survived – and thrived.

OXYGEN FROM WATER?

Yes, fish do get the oxygen they need from the water that they swim in. But they get these atoms of oxygen from *in between* the water molecules, not from *inside* the water molecules.

Water (H_2O) does in fact contain an atom of oxygen inside each molecule of water.

And plants do split up water to generate oxygen. Plants make the oxygen they release by this chemical reaction:

$$2H_2O + Energy \rightarrow 2H_2 + O_2$$

But fish did not evolve down that particular pathway. They don't break apart molecules of water to get at the atoms of oxygen from inside.

Instead, fish get their oxygen by 'filtering' out the occasional oxygen molecules that lurk in between the water molecules.

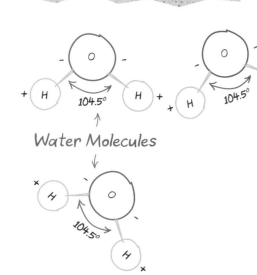

Water Molecules

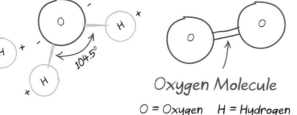

Oxygen Molecule

O = Oxygen   H = Hydrogen

You can see that there's an occasional molecule of oxygen (O_2) scattered around among the molecules of H_2O that make up water. These oxygen molecules are what the fish need, and what they have to separate out from the molecules of H_2O.

Sometimes, as in the terrible Australian Murray-Darling Rivers Fish Kills of 2019, there are not enough oxygen molecules in among the water molecules. In these cases, the fish can die of oxygen starvation.

WHAT KILLED A MILLION FISH IN THE MURRAY-DARLING RIVERS?

First, how do molecules of oxygen get *into* the water? Where do they come from?

These molecules might come from phytoplankton, tiny floating plants in the water. (Surprisingly, phytoplankton make about half or more of all the oxygen we humans breathe. So, half the oxygen in the atmosphere comes from the water, and half from the land.)

But the oxygen molecules do also soak in from the air immediately above the water. In this case, the oxygen gets in slowly if the water is still, but more rapidly when the water is turbulent and flowing.

Water can carry a maximum of 7–10 milligrams of oxygen in each litre.

However, the oxygen level in water can vary dramatically, dropping from high to so low that there's none left for the fish.

Here's a scenario. Start with water loaded with lots of oxygen, which also carries various 'organic' substances (fats, proteins, carbohydrates, etc). These organics are potential food for tiny living creatures, like bacteria. In the course of their normal existence (eating, digesting, growing, reproducing, etc), these bacteria also have to consume some of the oxygen in the water.

This usage of oxygen (by bacteria, algae, etc) is called the BOD, or Biochemical Oxygen Demand. It's usually measured in milligrams of oxygen per litre of water (mg/L). But the BOD is averaged out over a period of time.

Why? Well, when we humans eat a meal, it takes more than a second. It's the same for the bacteria living in water. A 'standard' period of time is usually considered by water-quality scientists to be five days. So when you see BOD_5 (mg/L), it refers to the number of milligrams of oxygen taken out of each litre of water, over a five-day period.

As an example, raw untreated sewage is full of 'delicious' nutrients for bacteria and algae. It can typically have a BOD_5 of 300–500 mg/L. That means to digest all the goodies in the sewage, the bacteria would need to consume 300–500 mg of oxygen in each litre of sewage. That's about 40 times more than the amount of oxygen actually present in the water. That's obviously a big problem.

It means the hungry bacteria would use up all the available oxygen in just a few hours, leaving nothing for other bacteria – or fish. The bacteria would die, and shortly after, so would the fish.

So, when bacteria treat the sewage in a biological sewage plant, you need huge numbers of bacteria, plus all the oxygen they need. This can be done by bubbling air through the sewage, or by surface aeration (churning the sewage around and around with big paddles). If you keep supplying oxygen to the bacteria, they can keep eating the sewage.

After good treatment, sewage can have its BOD_5 dropped from 500 mg/L to 5 mg/L.

Leading up to the multiple 2019 Murray-Darling rivers fish kills, huge amounts of algae died due to the temperature change. (Mind you, a healthy river system would not have a massive overgrowth of algae in the first place.) Dead algae to you and me, but food for bacteria. Of course – waste not, want not – the bacteria started eating the dead algae. The bacteria used up oxygen as they ate the algae, and there wasn't any oxygen left over for the fish.

Over a million fish died in a few days.

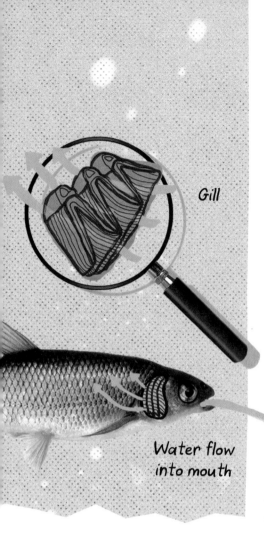

Gill

Water flow
into mouth

FISH GILL

Fish have been around for about 450 million years. That's plenty long enough to have evolved an organ to capture oxygen molecules from the water.

That organ is the 'gill' – their version of a lung. The gills are so important that the entire blood supply coming out of a fish's heart goes first to the gills, and only afterwards, to the rest of the body.

Gills are light feathery structures with lots of surface area. On one side is the ocean water, and on the other side is a very rich blood supply. Oxygen molecules are small enough to passively diffuse – or pass – from the water into the blood of the fish.

The fish open their mouth, suck in water, close their mouth, and then push the water out – through the gills. Gills are exceptionally good at exchanging gas (such as oxygen). The distance that the oxygen molecules have to traverse between the water and the bloodstream is a little smaller than 1 micrometre in tuna. Gills can remove between 50–90% of the oxygen present in water, as it passes through the gills in a single pass.

So, hooray, the fish can successfully transfer oxygen from the water they swim in into their bloodstream. This happens with both the salties and the freshies.

OSMOSIS 101 – CONCEPT OF 'GRADIENT'

But why do the salties swallow, and the freshies not?

This brings us to the strange concept of 'osmosis'. One definition of osmosis is *the tendency of water to move from a region of high water concentration to a region of low water concentration.*

So first, to understand osmosis, think of *water*, not of salt (or other particles). In osmosis we look at the concentration of water molecules, not the concentration of the 'particles' in between the water molecules.

You just have to think in a different way – *water*, not particles. Consider a litre of water, with a bit of salt in it. That litre contains a lot of water molecules, and a few salt molecules. In this setting, if it has a high concentration of water, it has a low concentration of salt. And vice versa.

The second thing you need to know to understand osmosis is that most 'things' spontaneously go downhill. First example: a ball will roll down a hill. The ball is going down the energy 'gradient' – from high potential energy at the top of the hill, to low potential energy at the bottom of the hill.

Second example: heat will go down a 'heat gradient'. Heat will go from a hot place (high thermal energy) to a cold place (low thermal energy).

And yes, the water concentration will go downhill, from high to low. This means the concentration will rise from low to high.

OSMOSIS 102 – CONCENTRATION GRADIENT

So now we're ready to work out *why* salties swallow and freshies don't. Let's start with a U-tube tank, half-filled up with fresh water. Next, we separate the two halves of the U-tube with a 'semi-permeable membrane'. In this case, 'semi-permeable' means that water molecules can get through this membrane, but other molecules cannot. (There are various reasons why these other molecules can't cross. They might be too big. They might carry an electrical charge. But the important thing is that they cannot get across, while water molecules can.)

Now let's tip in lots of common table salt (sodium chloride, or NaCl) on one side of the membrane and a little on the other side. The salt dissolves in the fresh water. This water now has a higher concentration of salt than does the other side.

Remember that 'stuff' tends to go downhill?

There's now a high concentration of salt on the left side of the U-tube, but a low concentration on the right.So, following the rule (that stuff goes downhill), the atoms that make up the salt 'want' to go to the right – to travel to the area of low concentration.

But they can't. The semi-permeable membrane won't let them across. In a fish, the semi-permeable membrane is in the skin and gills.

OMG, how does Nature fix this problem?

Easy. If the salt cannot go to the right, then the water (which *can* pass through the membrane) will go to the left. The water goes from a high concentration of water (100%) to a lower concentration of water (say 99.5%). The result is the same. The high concentration of salt molecules gets diluted to a lower concentration.

So this is why, in osmosis, you have to think of water moving *down* its concentration gradient.

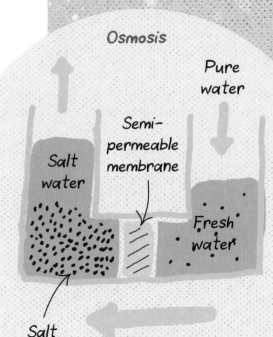

Osmosis

Pure water

Semi-permeable membrane

Salt water

Fresh water

Salt

Direction of water flow

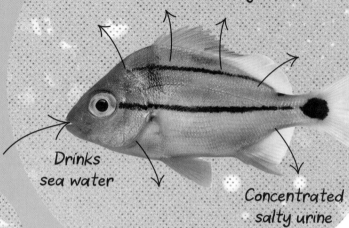

OSMOSIS 103 – SALTWATER FISH

It turns out that the level of salts in the salty oceans (around 1,000 mOsm/L), is higher than the level of salt in the flesh and blood of the saltwater fish (around 500 mOsm/L). So, thinking *water*, the concentration of water in the salty oceans is lower than in the saltie.

The salt in the oceans cannot cross the skin of the saltwater fish to go into the fish. But the water molecules can – and in the opposite direction.

So out go the water molecules, trying to dilute the entire salty oceans of the world. (This is an impossible task – fish are small, and the oceans are huge.)

This means that a saltwater fish is always getting rid of water (through its gills and skin).

The first thing it does is swallow all the water it can (even though it's salty). Each hour, it will swallow about 3–10 ml ocean water/kg of body weight. The second thing is that the saltie cuts right back on making urine. (After all, it doesn't need to lose more water.) In fact, because they don't need to make much urine, saltwater fish have relatively small kidneys.

OSMOSIS 104 – FRESHWATER FISH

It turns out that the level of salt in the freshwater lakes and rivers is close to zero (otherwise, it would be salty water, not fresh water). The level of salt inside a freshwater fish (about 250 mOsm/L) is higher than the level of salt in the water it swims in (around 1 mOsm/L).

The salt in the body of the fish cannot cross the gills and skin of the freshwater fish to go into the surrounding water. But the water molecules *can* cross the gills and skin – into the bloodstream of the fish.

So the water molecules from the river gush in, trying to dilute the internal saltiness of the fish.

This means a freshie has no need to actually *drink* water – because it's already getting a deluge via the gills and skin. In fact,

Saltiness?

Osmosis describes the movement of *water*. The water moves because of different concentrations.

We commonly describe the concentration of particles in the water by its 'osmolarity'. We measure osmolarity with a strange unit called a milli-osmole, usually shortened to 'mOsm'. So here are some numbers.

	mOsm/L
Fresh water	1
Freshwater fish	274
Saltwater fish	452
Sea water	1,050

the freshwater fish is at risk of swelling like a balloon and popping open from all this water. So, to get rid of this excess water, it has big kidneys to help get rid of lots and lots of urine.

(Try not to think too much about that next time you swim in a freshwater pond full of fish. Yes, fish wee in fresh water.)

So the freshie does two things: it does not swallow the water that comes into its mouth, and it makes huge amounts of diluted urine.

So, when it comes to the question of 'Do fish drink?', there's a two-part answer: salties swallow the sea, while freshies fatten and wee.

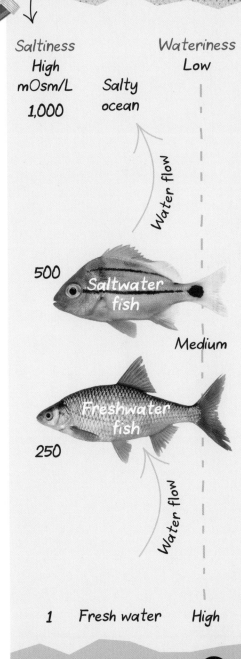

I've rounded off these numbers to make them easier to remember.

Saltiness
High
mOsm/L
1,000

Salty ocean

Wateriness
Low

Water flow

500

Saltwater fish

Medium

Freshwater fish

250

Water flow

1   Fresh water   High

Freshwater fish in fresh water

Water absorption through skin

Dilute urine

Try not to think too much about that next time you swim in a freshwater pond full of fish. Yes, fish wee in fresh water.

HUMAN FAULTS

It took me a long time to learn about 'life'. My first university degree was in physics and mathematics. As you would expect, it included absolutely nothing about any form of biological life – whether plant, fungus, bacteria, virus or animal. As far as I knew from my university studies, the human body was probably filled with some kind of chunky red salsa.

When the movie hero got shot in the shoulder by the bad guys, some of this red stuff would leak out. And that was kind of as much as I knew about the human body. (Okay, I had heard of bones, but what the liver, kidneys and spleen do? No idea.)

But a few years later, I ended up studying human physiology, and I was absolutely astonished by the sophisticated engineering that exists within our bodies – such as the amazing counter-current filtering mechanisms in the kidneys.

But the flipside was being astonished by the number of body parts that could do with a major upgrade. Believe it or not, there's a lot of bad engineering in the human body.

EYES WRONGLY PLUMBED

Take your eyes for example. They're each about the size of a golf ball.

Lining the back of each eye is a thin layer (0.3 mm) called the retina. Incoming light lands on the retina, and gets turned into electricity. This electricity gets partly processed in the retina, and then gets shunted into the optic nerve and sent to parts of the brain at the back of your skull. These parts then process the electricity to give you that full glorious wraparound 3D colour vision that we love so much.

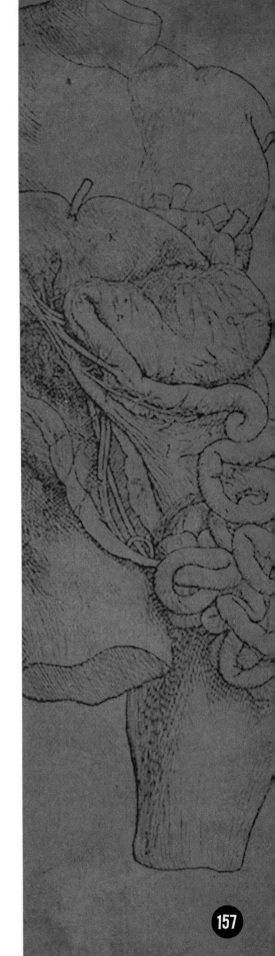

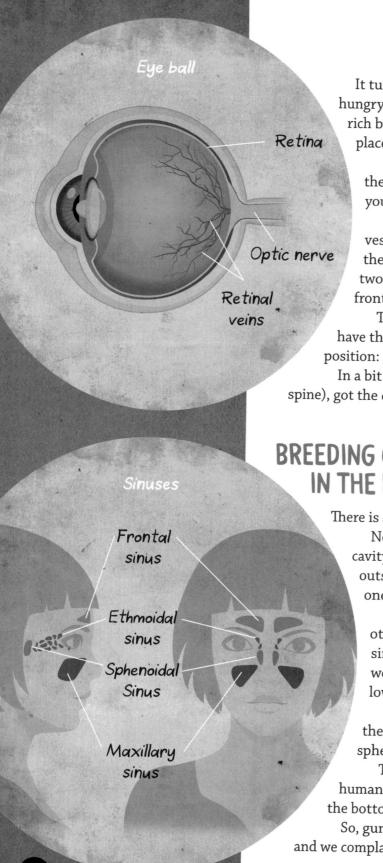

Eye ball

Retina

Optic nerve

Retinal veins

It turns out that the retina is one of the most oxygen-hungry tissues in the body. So, of course, it needs a very rich blood supply. And where are these blood vessels placed?

Right in *front* of the retina! This is crazy, because they block some of the incoming light! It's like putting your fingers in front of the lens of your camera.

Wouldn't it make more sense for these blood vessels to be behind the retina? Well, yes! And, in the animal kingdom, evolution has produced at least two versions of the eye – one with the blood vessels in front of the retina, and one with them behind.

The cephalopods (such as the octopus and squid) have the blood vessels for their retina in the logical position: *behind* the retina.

In a bit of bad luck, vertebrates like us (critters with a spine), got the crappy version.

BREEDING GROUND FOR INFECTION IN THE HEAD

Sinuses

Frontal sinus

Ethmoidal sinus

Sphenoidal Sinus

Maxillary sinus

There is a common condition called sinusitis.

Now, 'itis' means 'inflammation', while a 'sinus' is a cavity in the bone with just one single opening to the outside. So 'sinusitis' simply means inflammation of one or more sinuses.

As a result of this inflammation, 'gunk' and other debris can build up inside one of the several sinuses in your head. In an ideal world, this gunk would drain out easily through an opening at the lowest point in the sinus.

Humans have four pairs of major sinuses in the head – the frontal, maxillary, ethmoidal and sphenoidal.

The maxillary is the largest. But wait for it – in humans, the opening is near the top of the sinus, not the bottom. It's the same for the sphenoidal sinus.

So, gunk just sits there and gets infected and festers, and we complain.

BADLY LOCATED VOICE NERVES

As another example of bad engineering, look at the nerves that control the larynx. The larynx modulates the flow of air out of the lungs, and so gives us the ability to speak. The larynx gets its instructions via a pair of nerves at the top (the superior laryngeal nerves) and at the bottom (the recurrent laryngeal nerves). If the recurrent laryngeal nerves stop working, you can still speak, but only in a hoarse voice. Your words can be understood, but you can no longer speak and sing clearly and melodiously.

So, let's look at the nerves that control the bottom of the larynx. They start off (coming out on each side of your brain) as the two vagus nerves. They end up going all through your chest and gut, down to the anus. But along the way, each vagus nerve throws off a little branch to run along the bottom of the larynx.

Now the larynx itself is pretty close to the brain, in your neck. Sounds like a very straightforward engineering problem to solve: simply run the nerves that control the larynx on a nice short rung directly from the brain to the larynx?

Um, no. Instead, the nerves that control the bottom of your larynx take a ridiculously long and convoluted pathway *past* the larynx where they will ultimately end up. They continue down into your chest, underneath a major artery (the arch of the aorta on the left, or the subclavian artery on the right), and then all the way back uphill to your larynx.

Why? You can blame that messy wiring on our early vertebrate ancestors living in the ocean. Back then, a straight path from the brain to the gills zipped past the heart; and, back then, the heart was close to the brain. But we left the oceans to come onto the land some 400 million years ago. During this evolutionary transition from floating in sea water to walking on the land, the heart shifted to a new location. It moved, and as it moved, it dragged the nerves that controlled the larynx with it.

These nerves are the recurrent laryngeal nerves. 'Recurrent' means that the nerves take a convoluted pathway, and that they retrace their path. They go somewhere unnecessarily far away, then come back again. Unfortunately, these nerves go very close to the thyroid gland in the neck. This means that in thyroid surgery there's a risk of accidentally cutting these nerves. And, yes, this happens. The patient is left with a hoarse voice for the rest of their life.

You might think that I'm being too picky. The nerves that give fine and precise control over the voice are a recent addition.

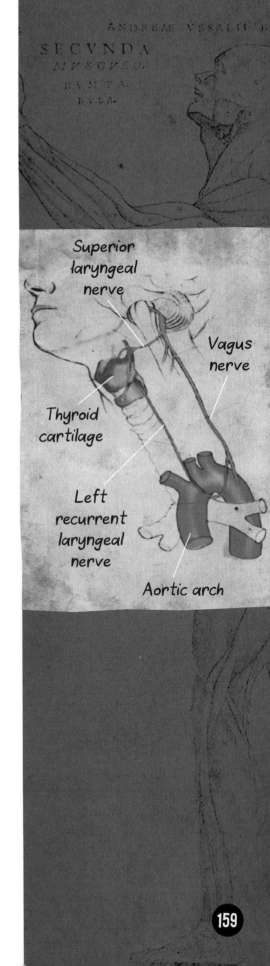

Superior laryngeal nerve

Vagus nerve

Thyroid cartilage

Left recurrent laryngeal nerve

Aortic arch

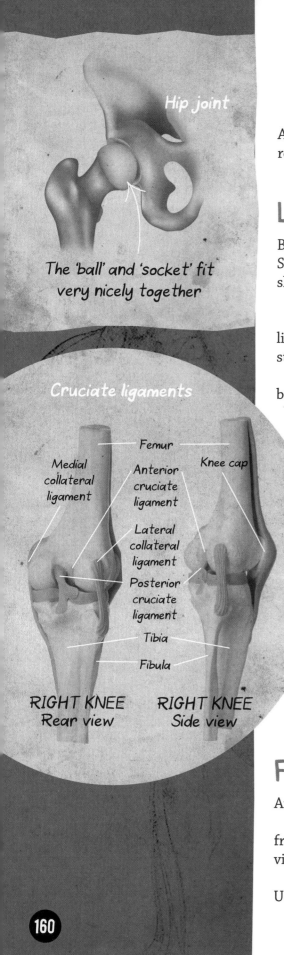

The 'ball' and 'socket' fit very nicely together

Cruciate ligaments

Femur

Medial collateral ligament

Anterior cruciate ligament

Knee cap

Lateral collateral ligament

Posterior cruciate ligament

Tibia

Fibula

RIGHT KNEE
Rear view

RIGHT KNEE
Side view

After all, we have had speech for probably less than a million years – really not that long to get the location just right.

LEG JOINTS

But skeletons have been around for hundreds of millions of years. Surely, you ask, evolution would have got something like the skeleton properly sorted out?

Again, yes and no!

Well, from an engineering point of view, the hip joint is a nice little joint. It's a ball-and-socket arrangement, and so is inherently stable. I like the hip joint.

But the knee joint looks like it was just tossed together. The bone above (the femur) doesn't sit in a very deep socket in the bone below. Instead, the knee joint surfaces are like shallow saucers sliding over each other. They're held from falling apart by strips of gaffer tape – okay, ligaments – on the left and the right, and the front and the back.

And there are even two strips of 'gaffer tape' coming straight through the middle of the knee joint, between the pair of shallow saucers. They run from the top to the bottom, and from front to back, in the shape of a cross. These are the constantly injured anterior cruciate ligament (ACL), and the posterior cruciate ligament (PCL).

The knee joint looks a mess. And from an engineering point of view, it is.

As you would know, knee injuries are all too common in so many sportspeople. If you're watching a footballer limping off a field, you'll probably hear the commentator say, '… probably done the ACL'. ACL injuries have finished many an athlete's career.

FRIENDLY FIRE FROM THE IMMUNE SYSTEM

And then there's the immune system.

The immune system is absolutely essential in protecting us from attack by external bad guys. These bad guys include bacteria, viruses and parasites.

But according to the National Institutes of Health in the United States, about 23.5 million Americans (that's about 7% of the population) suffer from some kind of autoimmune disease.

Now 'autoimmune' means that instead of your immune system defending you, it starts attacking you. Unfortunately, this is what happens in autoimmune diseases.

Autoimmune diseases include myasthenia gravis (where the muscles become progressively weaker), and better-known diseases such as Type 1 diabetes and hypothyroidism. Graves' disease (an autoimmune hyperthyroid disease) affects 0.5% of men and about 3% of women in the USA.

The immune system is devilishly complicated, so maybe it deserves some leeway. It would be hard to tune something as elaborately complex as the immune system to perfection. I guess it's no wonder that it mixes up 'us' and 'them'.

NOT OUR DNA AS WELL?

Unfortunately, the imperfections in our bodies go all the way down to our very core. There are flaws even in our DNA, the map for making our bodies.

Thanks to the forensic TV crime shows, we've all heard of DNA, and DNA testing. It turns out that about 8% of human DNA is made up of dead viruses that have tried to attack us in the past. Their dead carcasses have been in our DNA for hundreds of millions of years. They have been pointlessly passed down from our ancestors, and we will pointlessly pass them on to our descendants. In most cases, they are harmless. They are not particularly worrying – they just take up space in the DNA.

But what about DNA and cancer? Why isn't our DNA better at fighting cancers?

Humans have only one copy of an anti-cancer gene called p53 (see **WHY ELEPHANTS DON'T GET MORE CANCERS** on page 72). But elephants have 20 copies – and as a result, hardly ever get cancers. I wish we had a hundred copies.

FLAWS ARE US

I guess the lesson from all of this is that evolution doesn't have to be perfect – just good enough. So long as we live long enough to have babies, and bring them up, then evolution is happy.

And it's our flaws that make us wonderfully human, after all.

MISSING MASS
OF AN ATOM

In the science world, there's plenty of talk about the 95% of the Mass of the Universe that is 'missing'. The mysterious Dark Energy and Dark Matter are the 'missing bits' in the sense that – while we are very confident they exist – we just don't know what they're made of.

And in a nice act of symmetry, what do we find when we shift our focus from the Really Enormous Entire Universe to the Really Tiny Atoms? More mass that is 'missing'. About 91%!

First, some background.

The visible Universe is made from atoms. These atoms are made up of protons and neutrons, which in turn are made of quarks. (We can leave out electrons; they weigh hardly anything.) Quarks are the end of the line – they're the smallest possible particles that we have come across, so far.

So, talking about the mass of an atom, you'd think you could just add up all of the masses of the quarks – and bingo, you would have the answer?

But no way – this is Quantum Physics! Adding up the mass of the quarks gives you just 9% of the atom's measured mass. What?!

The remaining 91% is, bizarrely, 'frozen energy', made up of 'crazy' quantum and relativity stuff. Even harder to get your head around, it's made up of fast-moving stuff that has no mass!

Pretty wild, huh?

POKING ATOMS

Atoms are the tiny particles that make up the 5% of the Mass of the Universe that we do know a little about. Let's start with three slightly mind-bending factoids about atoms.

First: most of the atoms in the Universe are the two lightest atoms – hydrogen and helium.

Atoms at school

Currently in Australia, the concept of atoms is not officially taught until high school.

This seems to be related to a belief that young children can't handle abstract concepts. In turn, this idea seemed to evolve from the work of the psychologist Jean Piaget.

However, judging by what I've heard, teachers who do teach 'atoms' in Years 1 and 2 find that their students absorb this knowledge quite easily – in fact, the students love it!

Knowing about atoms actually makes it so much easier to understand the world around you. Kids are curious, always wondering about stuff. You need to know what atoms are, and how they behave, to explain everyday phenomena – such as why your swimming goggles don't mist up if you smear the inside with shampoo, where the carbon in the carbon dioxide that you breathe out comes from, and why it gets colder as you climb a mountain. (Each one of these would take several hundred words to explain, so I won't do it now.)

Mass of Matter

Almost all of the mass of our planet comes from the nuclei inside atoms. The electrons have hardly any mass – they're about 1,800 times lighter than protons and neutrons.

Their relative masses are:

Proton	1.673×10^{-27} kg	Relative Mass 1
Neutron	1.675×10^{-27} kg	Relative Mass 1
Electron	9.109×10^{-31} kg	Relative Mass 0.00055

Second: atoms are mostly empty space.

And, third – this is the new info – only about 9% of the mass of an atom comes from stuff that actually has mass.

In the early part of the 20th century, we were pretty sure that atoms existed – atoms in elements such as hydrogen, carbon, gold and so on. We also knew they had positive and negative charges 'somewhere' inside them. But we didn't know exactly where and how these charges were arranged.

So the scientists back then did their equivalent of the old 'Let's poke it with a stick to see what happens' technique. They threw one object at another object to see what happened when they collided.

Scientists such as Hans Geiger, Ernest Marsden and Ernest Rutherford started with gold foil – so thin it was only some 400 atoms thick. They then fired tiny charged particles (alpha particles, which are helium nuclei) at this gold foil. They hoped that the charged particles they were blasting at the gold foil might interact with the charged particles inside the atoms of gold. This would give them some clues about what was inside an atom.

Some of the charged particles they fired at the gold went straight through the foil, unchanged. No surprise – that probably meant there were spaces between the atoms in the gold foil, and/or there might be empty spaces inside the atom.

Some of the particles changed their path slightly. Again, no surprise – this probably meant that the tiny charged particles had collided with some component of the atoms of gold.

But, to the researchers' absolute surprise, some of the charged particles came straight back at them! Reflecting on this bizarre finding, Rutherford said, 'It was almost as incredible as if you fired a 15-inch shell at a piece of tissue paper and it came back and hit you'.

This result told us that atoms had a very small and very dense core that the charged particles were bouncing off. This core had both mass and charge. We now call this core the nucleus of the atom. (By the way, there was no real 'bouncing', because a nucleus is not a solid thing, but a fuzzy thing. The change in direction of the alpha particles was because of the charge of the nucleus.)

ATOMS 101

Experiments like this helped to develop our current model of the atom. An atom is mostly empty space, with a few tiny interesting bits in the middle and around the edges. And the nucleus (the bit in the middle) is very small compared with the size of the whole atom.

Think of a ball sitting all alone in the middle of a football field, with seats on the perimeter. The ball is the nucleus. And where the crowd sits are the electrons. But the electrons are not stationary – they're whizzing (or existing in a cloud) around the stadium.

The nucleus is where practically all of the mass of the atom is.

Most atoms have two types of particles in the nucleus – protons and neutrons. Protons are positively charged, while neutrons have no charge. (I remember this as 'positive proton' and 'neutral neutron'.)

Protons and neutrons have a very similar mass, but neutrons are slightly heavier. We have to go further down into the Tiny Land of Atoms to find 'mass'.

So far, on our journey down into the Land of the Tiny, we have gone from Matter (like gold) down to atoms, and down again to what atoms are made of (protons, neutrons and electrons). Let's keep diving even deeper.

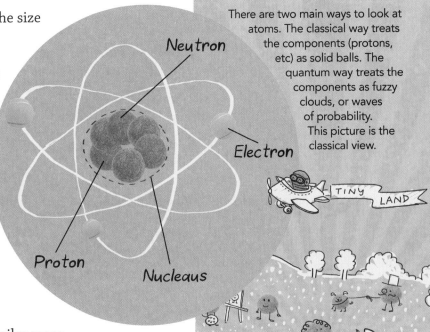

There are two main ways to look at atoms. The classical way treats the components (protons, etc) as solid balls. The quantum way treats the components as fuzzy clouds, or waves of probability. This picture is the classical view.

THE HIGGS FIELD

For most of us, a 'field' is where you have a picnic, a farmer grows a crop, or happy lambs gambol freely.

But for a physicist, a 'field' is a physical quantity that occupies space, and has energy. At each point in space and time, the field has a certain value.

There's an invisible magnetic field around a bar magnet. At any point near the magnet, this magnetic field has both a certain amount of 'strength' and a certain direction. If you sprinkle iron filings near the bar magnet, you can see

them get attracted to the magnet (the strength) and if you do it carefully, you can 'see' the direction of the magnetic field lines (by where the iron filings line up).

But some fields have only a value, not a direction. The Higgs Field is such a field, and it permeates the entire Universe.

What does it do? It gives mass only to fundamental particles, such as quarks. (Mass?!?!? This is hard to understand. But don't worry. If I could explain it in two sentences, it wouldn't have won a Nobel Prize.)

In any cubic metre anywhere in the Universe, the energy in the Higgs Field is equal to the amount of energy that our Sun would radiate in a thousand years. We currently have no way of getting access to that energy, but our children, or their children, might, as long as they do it very carefully ... (That kind of energy could evaporate the entire Earth into its constituent atoms. It wouldn't even notice the Earth was there. There wouldn't even be a burnt husk of the Earth left behind.)

Particle physics

I love everything about particle physics, except for the name.

What do you think of when I say 'astronomy'? Sitting around the campfire, on a cloudless and moonless night, with the Milky Way arching across the sky, and lots of meteors flashing by? Very nice.

But 'particle physics'? My mental image is of somebody dispiritedly staring at some particles of dirt.

No, let's call it something more impressive, such as 'fundamental cosmic physics', or 'high-energy physics' to represent the magnificence and grandeur of the Tiny Unknown Universe Down There.

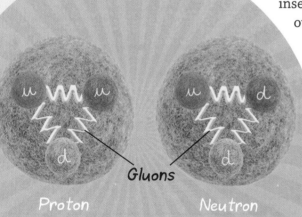

Gluons

Proton

A proton is made of two up quarks, and one down quark.

Neutron

A neutron is made of two down quarks, and one up quark.

HIGGS FIELD + FUNDAMENTAL PARTICLE → MASS

Atoms are made from three major 'particles' – electrons, protons and neutrons. But what are these made of?

To answer that, we have to get down to fundamental particles – which are the 'end of the line'. You can't go any smaller – as far as we know in 2019.

An electron is straightforward. It's a fundamental particle. An electron is not made of any smaller particles.

But protons and neutrons are not fundamental particles. They're made of smaller particles called 'quarks'. Quarks are fundamental particles.

There are six types of quarks – with cute names such as Top, Bottom, Strange, Charm, Up and Down. (And of course, there are anti-quarks – but let's leave them for another time.)

To keep it simple, think of a proton as being made of three quarks – two Up quarks and one Down quark. (Neutrons have two Down quarks and one Up quark.) The quarks are held together by massless particles called gluons. We know that these quarks have a certain amount of mass. This is reasonable – you would expect this. In 2012, we solved a long-standing and very important mystery about one aspect of Matter.

Now, Matter has various properties. One property is volume. If you fill a bathtub with water right to the very brim, and slowly insert your foot and then the rest of your leg, the water overflows. We understand volume pretty well.

Another property is optical. When light falls on your body, some is absorbed and some is reflected. When the reflected light lands on the cones in my retina, I can see you. We understand the optical properties of Matter moderately well.

But we didn't understand the property of Matter called mass. We had no idea where mass came from. Then, in 2012, we finally discovered that fundamental particles appear to get their mass by interacting with the 'Higgs Field'. Scientists worked this out using the Large Hadron Collider, on the border of Switzerland and France.

They proved that the Higgs Particle existed, and therefore there had to be a Higgs Field. This Higgs Field was first proposed in the 1960s. This was such a big discovery that it won a Nobel Prize.

Now here's something weird. The Higgs Field gives mass only to fundamental particles – such as quarks (and electrons). (As an aside, the Higgs Field does not interact with photons – which is why photons appear to have no mass.)

Let's keep it simple to start with, and look only at the proton, not the whole atom. When you do the numbers, the mass of the three quarks inside the proton adds up to only 9% of the total mass of the proton. So where does the other 91% of the proton's mass come from? Quantum and relativity, that's where!

THE MISSING 91%

The key to finding this missing 91% is to recognise that 'mass' and 'energy' are the same thing (but under different conditions).

You might have heard of Einstein's famous equation:

'E' stands for energy, 'm' stands for mass, while 'c' is the speed of light (300,000 km/sec). If you can turn some mass into energy, this equation tells you exactly how much energy you get.

But on a deeper level, this equation tells us that energy and mass are different sides of the same coin. You can think of energy as mass that has been liberated and set free to run around. On the other hand, mass is just energy that has calmed down enough to stay in one spot. Mass is just coagulated, or frozen, energy.

The bottom line is that energy and mass are interchangeable. This gives us some clues as to where all that hidden mass of a proton might be.

The three quarks inside the proton don't just sit there, totally stationary. They jiggle around like mad. And remember that the proton has three quarks inside it, accounting for 9% of its mass.

But there's something holding these quarks together inside the proton, stopping them from escaping. This 'something' is part of the other 91% of the mass of a proton. (And this applies to the neutron as well.)

These quarks are held together by the action of something called the Strong Force. (And rather predictably, it's the strongest of the four known forces in fundamental physics.)

The Strong Force acts via things we call 'gluons' (which kind of glue the quarks together). A proton's three quarks are held together by a tricky dance between these gluons.

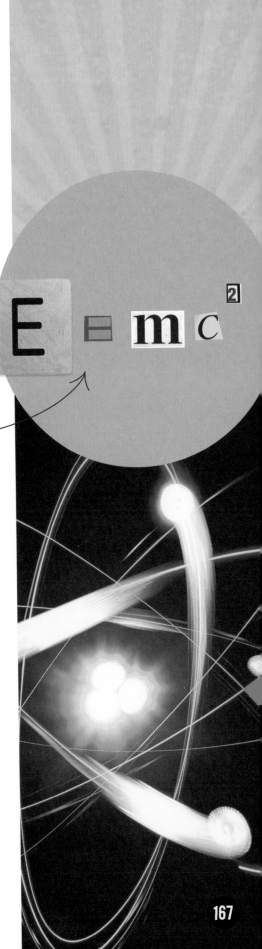

$$E = mc^2$$

Echoes of the Big Bang

Back at the Big Bang and for a little while after, the energy density of the Universe was enormous. Since then, the Universe has expanded and cooled, and most of this energy density has disappeared.

But today, a tiny amount of this ancient energy is carried by quarks and gluons – which are trapped inside protons and neutrons, which in turn are trapped inside atoms, and us.

So, it's rather nice to know that we carry some energy remnants of the Big Bang inside us.

Proton

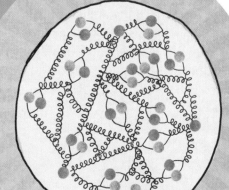

The inside of the proton has a dynamic maelstrom of many quarks (green dots) and anti-quarks (blue dots) winking in and out of existence, and being held together by gluons (springs). But averaged out over time, there are just three quarks being held together by three gluons.

MISSING MASS IS MOVEMENT!

Gluons have another unusual property: they spontaneously turn into quarks, and then back into gluons – and they do this all the time! (Yes, I know that 'quarks' are fundamental particles, and therefore aren't made of smaller particles. But, I don't know enough Lattice Quantum Chromodynamics to understand this – my bad!)

This is just one of the many weird things that very, very small things do – and these interactions are described by the mind-bending laws of quantum mechanics.

Apparently, in this seemingly magical Quantum Land, you can create pairs of particles out of nothing, and turn them back into nothing – providing you do it within a short enough time! (It's like teenagers borrowing their parents' car. If they can get it back into the garage before the parents notice, then it never happened. 😉)

So, gluons can actually decay into a pair of quarks (specifically, a quark and an anti-quark). But because these two are opposites, they 'add' up to nothing. But also because they are opposites, they can annihilate each other and turn back into a gluon – which is something. Does that make your head hurt?

So, according to all this, what does the inside of a proton look like then?

Well, we already know there are three quarks. But there are also gluons streaming between these three quarks. And there are also gluons spontaneously turning into pairs of quarks, and then back into gluons again. It's an airy fairy mess.

INSIDE A PROTON

So the inside of a proton is a seething quantum soup of lots of gluons and many pairs of quarks, winking into and out of existence. Sure, averaged out over time, there are only three quarks – two Ups and one Down. But, in reality, there are dozens of quarks popping in and out of existence all the time. The inside of the proton is actually a dynamic system, with lots of quantum fluctuation stuff happening inside.

All of this activity is what gives a proton (and you) the majority of its energy (which is the same as mass).

This is where a proton's missing mass comes from!

As you can imagine, it's incredibly difficult to mathematically calculate exactly where all of this energy (or mass) is inside a proton. But physicists have come up with Lattice Quantum Chromodynamics to try to do just that.

According to their calculations, 9% of the mass of the proton comes from the actual mass of the three quarks. This comes from the Higgs Field interacting with quarks (which are fundamental particles).

Next, 32% of the mass of the proton comes from the energy of quarks moving and bouncing around inside the proton.

About 36% of the mass of a proton comes from gluons – these particles that have no mass of their own. Their job (holding quarks together) involves a lot of energy, and this energy then appears as mass.

The final 23% of the mass of the proton comes from weird quantum effects due to quarks and gluons interacting with each other.

This is us – mostly condensed energy!

BACK IN THE REAL WORLD

So, why do we care about all this?

Well, first, understanding what makes up the physical world around us is a reasonably important project in itself. I love pure research.

And you don't know when pure research will pay off. About half a century ago, Stephen Hawking proposed that very low-mass black holes (if they existed) would give off radiation as they evaporated. The search for that radiation gave the world Wi-Fi. To cut a long story short, CSIRO invented the special technology needed to hunt for the radiation. Though CSIRO never found the radiation, decades later that technology was the basis for Wi-Fi, and it earned Australia one billion dollars in royalty payments. Who knew, back in 1972, that this would happen?

As another example, proton beam therapy did not come into existence because a medical doctor wanted a new tool for cancer therapy. It's here because, in the early 1900s, Rutherford wanted to understand how atoms work.

But second, I personally find it useful to know that while chocolate – like the rest of our material world – feels reasonably solid, it's mostly empty space. Much of its mass comes from ghostly, ethereal interactions. So, that leaves me feeling a little less guilty when I dig in for a second helping.

Feynman on atoms

Richard Feynman was one of the true geniuses of the 20th century. He won a Nobel Prize in physics. Read his books *The Feynman Lectures on Physics* and *Surely You're Joking, Mr Feynman!*, or watch his videos on YouTube. He speaks in plain language, so he's easy to understand.

He once wondered about what was the single most important piece of knowledge that should be passed to future generations, if all our current knowledge was lost. He thought it should be about atoms. He wrote that 'all things are made of atoms – little particles that move around in perpetual motion, attracting each other when they are a little distance apart, but repelling upon being squeezed into one another'.

ROGUE PLANETS

We know that there are eight major planets in our Solar System. They all orbit the Sun in roughly circular orbits.

Well, how's this for weird?

Imagine a planet that does not 'belong' to a solar system. Instead, just like the 300 billion stars in our galaxy, the Milky Way, it's in some kind of orbit around the centre of the galaxy. (Actually, to be pedantically precise, we all orbit the common centre of gravity of all the mass in the Milky Way.) Generally speaking, this planet roams freely through interstellar space. Very rarely, it might even be captured by a star moving in roughly the same direction and become part of a new solar system.

Welcome to the 'rogue planet'. And according to the latest 'guesstimates', there could be tens of billions of rogue planets in the Milky Way.

I love the idea of an intrepid planet blazing its own trail through the void of space.

I love the idea of an intrepid planet blazing its own trail through the void of space.

ROGUE PLANET 101

A rogue planet is not gravitationally bound to any particular star. It also goes under the name of orphan planet, wandering planet, starless planet, nomad planet and interstellar planet.

It's still early days for this research. So far, we seem to have discovered about 20 or so of these rogue planets. Most of the candidates are single rogue planets. In 2006, we discovered two planets in orbit around each other, with no star nearby. These planets were seven and 14 times the mass of Jupiter.

One rogue planet has been discovered as close as 63 light years from our Solar System.

Part of the problem with this research is that rogue planets are fairly hard to find. This is because they don't emit any visible light. In addition, they're so small, they reflect hardly any light (and there's probably no star nearby to shine on them, anyway).

But some of them do emit infrared light (which is just heat), which can be detected and so we can find them. But then sometimes we don't know whether the object is a very big rogue planet, or a very small brown dwarf (a star that wasn't massive enough to ignite).

Probably the main method used to find rogue planets is 'microlensing', a phenomenon that lets us study distant objects that emit little or no light. 'Lensing' refers to the gravitational bending of light around an intervening object between the distant object and the observer. And when lenses magnify, the image is brighter.

LENSING, WITH PRETTY PICTURES

Over on the right of the illustration below, you can see one of our space telescopes, Chandra. It has a 1.2-m detector.

Over on the left, you can see a quasar – a very, very bright object at the centre of a very distant galaxy.

Real quasar

1

How to make a rogue planet

There are a few main theoretical paths for planets to leave their host solar system.

In the first case, an external star passes close by and interacts gravitationally with the planets inside another solar system. One or more planets can be nudged and tossed out of its orbit. This would usually, but not always, happen in the early days of the solar system being formed.

In the second case, the planets within the solar system interact with each other via gravity – and again, one of the planets is expelled.

A third pathway is when a star – in the centre of its own retinue of planets – in its later life loses a huge amount of mass as it evolves into a red giant. Under certain conditions, this mass being expelled from the star could thrust a planet out into interstellar space.

The fourth possible pathway again can happen in the early days, when a solar system is being formed. Over time, planets naturally grow bigger, using gravity to suck in material from the so-called 'protoplanetary disk', a disk of dense gas and dust rotating around a young, newly formed star. It's possible that this disk can be disturbed by the gravity of another object in the crowded environment of a stellar nursery. In this case, some big fragments might be ejected, and these could then coalesce to form a planet.

2

Real quasar

View from Chandra

3

Real quasar

Galaxy

4

Real quasar

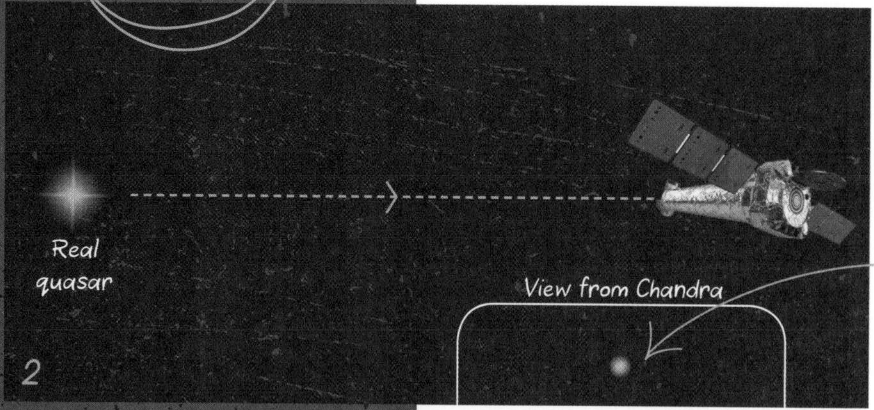

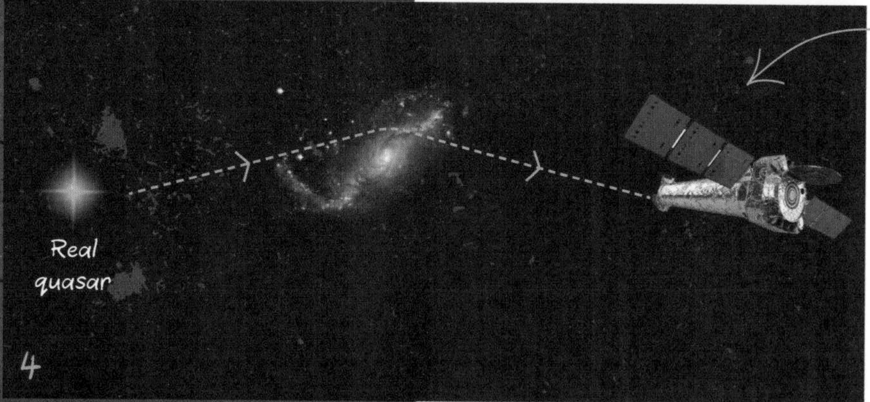

The light from this quasar spreads out in all directions. A few of the beams of light head in exactly the right direction to land on the detector inside the Chandra telescope. And at the bottom right of diagram 2, you can see that with this detector, we now capture a faint image of the distant quasar. It's faint because we are capturing only the faint amount of light that travels in a straight line to land on that relatively small 1.2-m detector.

Now, let's throw in a galaxy (a foreground object) in between the quasar and the Chandra telescope. This galaxy is not exactly in between the quasar and us, but pretty close to the straight line joining the quasar and us.

Many beams of light head past the intervening galaxy, on paths well away from the Chandra telescope. Normally, they would miss our telescope. But, thanks to the mass of the galaxy, some of these many beams are bent by exactly the right amount to go down the barrel of the Chandra telescope.

Chandra catches these beams of light if they end up in the telescope's barrel. It doesn't matter if their pathways are bent or straight. But when the light from the quasar curves around the galaxy before travelling a final straight section from the galaxy to Chandra, then a straight line backwards from Chandra seems to go to a point in space above the actual real quasar (see diagram 5). And so, the mirror on the Chandra registers two images – one faint 'direct' image, and the other, a brighter virtual image of the same quasar, but shifted a bit to the right.

Because there's a whole bunch of light beams that are all 'bent' towards our telescope, the resulting virtual image will actually be brighter than the one faint image coming directly from the quasar.

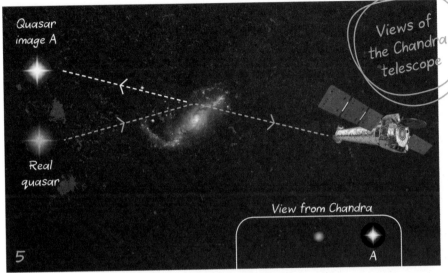

Views of the Chandra telescope

There are also other beams of light leaving the quasar to the left of the intervening galaxy, so we get two images (diagram 6).

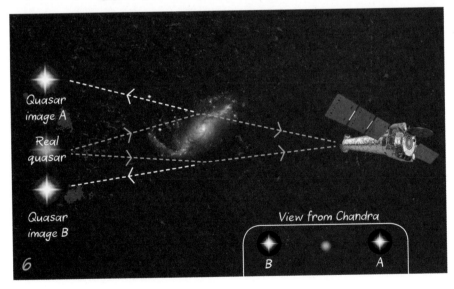

This second virtual image, again brighter than the original direct image, is shifted to the left of the real quasar.

So 'gravitational lensing' happens when we have a massive body (the foreground object) between us and the distant bright object (the background object).

Colliding rogue planets?

Could these rogue planets ever run into anything?

Yes-ish, according to the 2019 simulation of 2,522 planets orbiting 500 stars for 10 million years. The computer simulation shows that over that 10 million years, about 10% of the planets changed the paths of their orbits quite dramatically.

About 3% of the planets actually smashed into their hot star, while 0.6% collided with another planet in that solar system.

But don't worry. The chances of a disaster-movie scenario, of a rogue planet suddenly coming into view next year out past Pluto and slamming into Earth, are vanishingly small.

MICROLENSING

'Microlensing' is when the brightness of a star suddenly increases, because a non-massive object has briefly passed between the star and us.

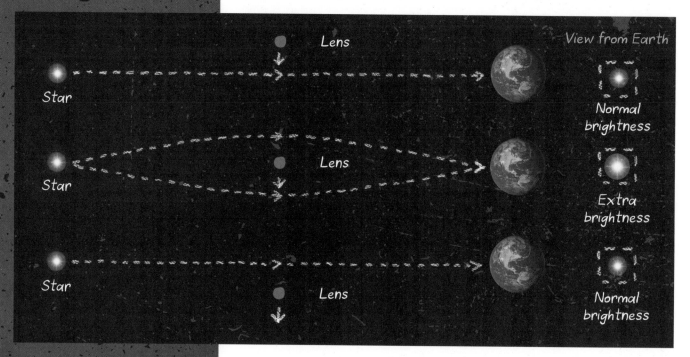

In the case of microlensing, the non-massive object does not have enough mass to bend space-time enough to form sharp images. So instead, we get a simple and short-lived brightening of the distant object.

One 2011 study looked at 50 million stars in our Milky Way galaxy, using the 1.8-m MOA-II telescope at New Zealand's Mount John Observatory. They observed all these stars, at least once per hour for a whole two years. They found 474 incidences of microlensing. Ten of these incidents lasted for less than two days. They were brief enough to be caused by a planet roughly the size of Jupiter. But five of these ten cases were probably orbiting at a great distance from their star. This left us with five where there was no star nearby.

Look at how quickly the numbers drop – 50 million stars, 474 microlensing incidents, and only five possible rogue planets. Such a small final number makes it hard to extrapolate the total number of rogue planets in the Milky Way. On the other hand, we have looked only for a few years, not centuries, and not millions of years.

When the astronomers took this all into account, they came

LENSING THEORY

'Lensing' in astronomy means that when a foreground object (like a planet, star or galaxy) passes in front of a background object that gives off light (like a star or galaxy), then that foreground object can act like a 'lens'. It can make that background object temporarily brighter. Depending on the geometry, that 'lens' can also give us two separate images of that object, or even a complete ring.

Welcome to yet another phenomenon that goes back to Einstein back in 1915 in his Theory of General Relativity.

Albert Einstein predicted that 'lensing' could happen, way back in 1912 – before the publication of his Theory of General Relativity (which is actually a Theory of Gravity). He was thinking that an object between a distant galaxy and us could act like a lens – it could 'bend' the light of the distant galaxy.

In 1924, the Russian physicist Orest Khvolson (or Chwolson) reckoned that if the distant source, the 'lens' in between and the observer were in a perfectly straight line, a ring of light could be seen about the intervening lens. Today, this is called an Einstein Ring, or an Einstein-Chwolson Ring. But this ring happens ONLY when all three are in a straight line. That's very uncommon. What is much more common is that they are not perfectly aligned, so an observer will see two images. In 1933, the Swiss astronomer Fritz Zwicky predicted that these two images would one day be seen.

(By the way, Zwicky invented a special Physics insult. When he called somebody 'a spherical bastard', he meant that they were a 'bastard', no matter which way you looked at them.)

Since then, we have seen many Einstein Rings and sets of two images of a distant galaxy – when there has been a body (an effective lens) in between the source and us.

How does this 'lensing' thingy happen?

The first thing to know is that the gravity of an object will bend space-time around it. If the object is a big, massive galaxy, it will bend space-time a lot. But if the object is a small, not-massive planet, it will bend space only a little bit.

The second thing to know is that light will follow this bent (or curved) space-time.

up with a rough 'guesstimate' of billions of possible rogue planets in the Milky Way.

But there are many other causes that can make stars get brighter. About 2% of the stars in the Universe change their brightness anyway, on a regular cycle. And of course, there are all kinds of other events (novas, supernovas, etc) that make stars change their brightness.

HOW MANY ROGUE PLANETS?

One big question is how many free-roaming planets there are in our galaxy.

One estimate in 2019 suggests that, in our galaxy, there could be some 50 billion of them. This finding came out of the University of Leiden in the Netherlands.

Our galaxy, the Milky Way, has some 7,000 regions called stellar

View of the Orion Nebula – a well-known region of star formation – via the Hubble Space Telescope. The Trapezium star cluster is the bright area just left of centre. It contains about 2,000 known stars, but there may be more as well. It is a young open cluster where the stars are all roughly the same age.

nurseries – this is where stars are born. One of these regions is the Trapezium Star Cluster, some 1,300 light years away in the Orion Nebula, in the direction of the constellation, Orion.

The Trapezium Star Cluster has some 1,500 stars. The astronomers set up a computer simulation to mimic this stellar nursery. It gave 500 of these stars simulated planets – four planets, five planets or six planets each. The planets came in many different masses (from 0.8% of the mass of Jupiter, to 130 times the mass of Jupiter, which is more of a red dwarf star and less of a planet). Overall, there was a total of 2,522 planets orbiting their mother stars.

The astronomers then set their planetary computer modelling simulation running. They started with the star (and planets) being 1 million years old, and ran the clock for 10 million years. By the end of that time, in this computer simulation, about 14% of those planets – 357 of them – had run away from their host stars to become free-roaming planets. Some 282 of these free-roaming planets totally left the Trapezium star cluster. This left 75 planets roaming along with this star cluster, but not bound to any individual star.

Following this mathematics further, over a longer period of time, the astronomers reckon that about one-quarter of the stars in the Milky Way have lost at least one planet. There 'could' be 50 billion planets meandering freely in our galaxy alone.

Closer view of the Trapezium Star Cluster in the Orion Nebula (bright stars near centre of photo).

OUR SOLAR SYSTEM?

And what about our own Solar System? Is there a planet beyond our Solar System (an exoplanet) that originated from within it? Has Earth lost one of its siblings?

For starters, we could have lost one of our planets in the early days of our Solar System's formation. After all, the most commonly found size of planet that we have discovered in our searches for exoplanets has been something with a size somewhere between Earth and Neptune.

DEAD OR ALIVE

At first glance, you might think that one of these planets roaming very, very distant from any nearby star would be cold, frozen and without an atmosphere or life. But anything is 'possible' in the Land of Theory.

For example, if a rogue planet could leave its solar system with enough hydrogen or helium in its atmosphere, it could keep warm even while being very distant from any stars.

Consider an Earth-sized planet, being ejected from its parent solar system very soon after being formed. Our Sun puts out about 40% visible light, about 50% heat, and neither of these has enough energy to knock electrons off atoms. But sunlight also carries about 10% ultraviolet. Ultraviolet light has enough energy to damage some atoms (by knocking off their electrons), and to cause skin cancer in human flesh. The ultraviolet light has much more energy than visible or infrared light, so that energy can be transferred to any atoms that it runs into. So when a star gets thrown out of its orbit when it's still young, there hasn't been enough time for ultraviolet light from its mother star to energise the hydrogen and helium atoms in the atmosphere, and boot these atoms out into space. Even a small planet the size of Earth would have enough gravity to stop hydrogen and helium from escaping its atmosphere.

Let's push things a bit further. Let's assume a special case, where the Earth-sized planet had an atmospheric pressure of 1,000 atmospheres of hydrogen (that's huge). In this rather specific case, the geothermal energy released from the decay of radioactive elements in the core could maintain the surface temperature above the melting point of ice.

It would be theoretically possible for this dark planet, far from any star, to have oceans of liquid water. If the planet was geologically active, it is just possible that hydrothermal vents could provide energy for life.

But this is all just speculation (or arm-waving, as the physicists call it) ...

This is one very soft hint that we might be missing a sibling. After all, we don't have a planet of that size in our Solar System. Was there such a planet ejected in the early days of our Solar System?

There are two more hints that our Solar System might have had a strong encounter with other stars in the early days.

First, there's a dwarf planet, Sedna, way out in our Solar System. How did it get there? An accidental side effect of a planet being ejected into interstellar space?

Second, there's another 'asteroid belt' out past Neptune, called the Kuiper Belt. The problem is that the Kuiper Belt is very scattered. Why? An accidental side effect of a planet being ejected into interstellar space?

And there may be some fireworks in the future for our Solar System. Some mathematicians specialise in planetary orbits and chaos theory. Some of them claim that over a long time period of hundreds of millions of years, there is a 5% chance that Mercury could be propelled into the Sun.

Maybe that's a fitting place for the messenger of the gods to end up.

END NOTES

Why is Wombat Poo Cube-Shaped?

'Why do wombats do cubic poos?', Louise Gentle, *Australian Geographic*, 18 April 2016, https://www.australiangeographic.com.au/topics/wildlife/2016/04/why-do-wombats-do-cube-shaped-poos/

'Wombats are the only animal whose poop is a cube. Here's how they do it', Laurel Hamers, 19 November 2018, *Science News*, https://www.sciencenews.org/article/how-wombats-poop-cubes

'Scientists unravel secret of cube-shaped wombat faeces', Ian Sample, *The Guardian*, 19 November 2018, https://www.theguardian.com/science/2018/nov/18/scientists-unravel-secret-of-cube-shaped-wombat-faeces

'Why is wombat poop cube-shaped?' Tik Root, *National Geographic*, 19 November 2018, https://www.nationalgeographic.com/animals/2018/11/wombat-poop-cube-why-is-it-square-shaped/?cmpid=org=ngp::mc=social::src=twitter::cmp=editorial::a%E2%80%A6

'Wombat poop: Scientists have finally discovered why it's cubed', Helen Regan, CNN, 20 November 2018, https://edition.cnn.com/2018/11/19/australia/wombat-cube-poo-intl/index.html

Closest Planet to Earth – And the Winner is ...

'Sugar, outdoors play and planets', BBC Radio 4 Podcast, 11 January 2019, https://www.bbc.co.uk/programmes/p06xvkjc

Genetic Testing Saves Babies

'Rapid whole-genome sequencing decreases infant morbidity and cost of hospitalization', Lauge Farnaes, et al, *npj Genomic Medicine*, 4 April 2018, https://www.nature.com/articles/s41525-018-0049-4

'Rapid Paediatric Sequencing (RaPS): Comprehensive real-life workflow for rapid diagnosis of critically ill children', Lamia Mestek-Boukhibar, *Journal of Medical Genetics*, Vol. 55, pp 721–728, https://jmg.bmj.com/content/55/11/721

'Hype becomes reality', Editorial, *New Scientist*, 31 March 2018, p 5.

'Genomics saves lives', Clare Wilson, *New Scientist*, 31 March 2018, pp 6, 7.

How Do Birds Sleep?

'Do birds sleep in flight?', Niels C. Rattenborg, *Naturwissenschaften*, September 2006, pp 413–425.

'Forget the nest: Where birds really go to sleep', Nicholas Lund, *Australian Financial Review*, 7 February 2014, https://www.afr.com/news/forget-the-nest-where-birds-really-go-to-sleep-20140207-ixuha

Ships & Ocean-Level Rise

'Ships: How much would the sea level fall if every ship were removed all at once from the Earth's waters?', Michael Toje, https://what-if.xkcd.com/33/

'How fast is sea level rising?', The Royal Society, https://royalsociety.org/topics-policy/projects/climate-change-evidence-causes/question-14/

'Global sea level rise rate speeding up, 25 years of satellite data confirms', by Genelle Weule, 13 February 2018, ABC, https://www.abc.net.au/news/science/2018-02-13/25-years-of-satellite-data-confirms-global-sea-level-rise-rate/9416570

'Four decades of Antarctic ice sheet mass balance from 1979–2017', Eric Rignot et al., *Proceedings of the National Academy of Sciences*, Vol. 116, No. 4; 22 January 2019.

'Here's how much ice Antarctica is losing – it's a lot', Andrea Thompson, *The Scientific American*, 30 January 2019, https://www.scientificamerican.com/article/heres-how-much-ice-antarctica-is-losing-mdash-its-a-lot1/?utm_source=newsletter&utm_medium=email&utm_campaign=sustainability&utm_content=link&utm_term=2019-01-31_featured-this-week&spMailingID=58360602&spUserID=NTM5NzI1MjE3NgS2&spJobID=1563995893&spReportId=MTU2Mzk5NTg5MwS2

Hole Phobia?

'Fear of holes', Geoff G. Cole and Arnold J. Wilkins, *Psychological Science*, 27 August 2013, pp 1980–1985.

'Disgusting clusters: Trypophobia as an overgeneralised disease avoidance response', Tom R. Kupfer and An T.D. Le, *Cognition and Emotion*, 6 July 2017), Vol. 32, No. 4, pp 729–741.

'Pupillometry reveals the physiological underpinnings of the aversion to holes', Vladislav Ayzenberg et al, *PeerJ*, 4 January 2018, DOI 10.7717/peerj.4185.

'Trypophobia and the science of disgust', Tim Wallace, *Cosmos*, 3 April 2018.

'Tryphobia 101: A beginner's guide', Nuna Alberts, *Everyday Health*, 31 July 2018, https://www.everydayhealth.com/trypophobia-101-beginners-guide/

'New Nokia phone with five camera lenses is terrifying people who suffer from fear of HOLES', Joe Pinkstone, MailOnLine, 26 February 2019, https://www.dailymail.co.uk/sciencetech/article-6742381/New-Nokia-phone-five-camera-lenses-terrifying-people-suffer-fear-HOLES.html#reader-comments

Big Brother Internet?

'Arkansas judge drops murder charge in Amazon Echo case', Nicole Chavez, CNN, 2 December 2017, https://edition.cnn.com/2017/11/30/us/amazon-echo-arkansas-murder-case-dismissed/index.html

'Hey, Alexa, what can you hear? And what will you do with it?', Sapna Maheshwari, *The New York Times*, 31 March 2018, https://www.nytimes.com/2018/03/31/business/media/amazon-google-privacy-digital-assistants.html?action=click&module=Intentional&pgtype=Article

'Alexa and Siri can hear this hidden command. You can't', Craig S. Smith, *The New York Times*, 10 May 2018, https://www.nytimes.com/2018/05/10/technology/alexa-siri-hidden-command-audio-attacks.html?emc=edit_mbau_20180510&nl=&nlid=1832479120180510&te=1

'Siri, Alexa and Google Assistant can be controlled by inaudible commands', Jeremy Horowitz, *VentureBeat*, 10 May 2018, https://venturebeat.com/2018/05/10/siri-alexa-and-google-assistant-can-be-controlled-by-inaudible-commands/

Coral Spawning

'Mass spawning in tropical coral reefs', Peter L. Harrison et al., *Science*, 16 March, 1984, pp 1186–1189.

'Sexy corals keep "eye" on moon, scientists say', William J. Broad, *The New York Times*, 19 October 2007, https://www.nytimes.com/2007/10/19/science/19coral.html

'Light-responsive cryptochromes from a simple multicellular animal, the coral *Acropora millepora*', O. Levy et al., *Science*, 19 October 2007, pp 467–470.

'Sexual reproduction of scleractinian corals', Peter L. Harrison, Z. Dubinsky and N. Stambler (eds), *Coral Reefs: An Ecosystem in Transition*, Springer, Netherlands, 2011, pp 59–85, DOI: 10.1007/978–94–007–0114–4_6.

'Coral spawning in the Gulf of Oman and relationship to latitudinal variation spawning season in the northwest Indian Ocean', E.J. Howells et al., *Scientific Reports*, 4: 7484, 15 December 2015, DOI: 10.1038/srep07484.

'That sinking feeling: Suspended sediments can prevent the ascent of coral egg bundles', Gerard F. Ricardo et al., *Scientific Reports*, 6: 21567, 22 February 2016, DOI: 10.1038/srep21567.

'In secrets of coral spawning, hope for endangered reefs', William J. Broad, *The New York Times*, 20 June 2016, https://www.nytimes.com/2016/06/21/science/coral-reproduction-survival.html?smid=tw-nytimesscience&smtyp=cur&_r=0

'The Great Barrier Reef's coral bleaching event is the canary in the coalmine for all the world's oceans', Cayla Dengate, *Huffington Post*, 3 August 2016, https://www.huffingtonpost.com.au/2016/04/13/coral-bleaching-great-barrier-reef_n_9657864.html

'Coral spawning less in northern reef, scientists say', Anna MacDonald, ABC News, 21 November 2016, https://www.abc.net.au/news/2016–11–21/coral-spawning-annual-eggs-sperm-coral-bleaching/8040900

'Coral spawning: a rare natural wonder', Kara Murphy and Angela Heathcote, *Australian Geographic*, 7 December, 2017, https://www.australiangeographic.com.au/topics/wildlife/2017/11/coral-spawning-a-rare-natural-wonder/

'Global warming impairs stock-recruitment dynamics of corals', Terry P. Hughes et al., *Nature*, 3 April 2019, https://www.nature.com/articles/s41586–019–1081-y

'Coral spawning drop raises fears for Reef's resilience', by Peter Hannan, *The Sydney Morning Herald*, 4 April 2019, pp 8, 9.

Tongue Taste Map

'Zur Psychophysik des Geschmackssinnes' David P. Hänig, *Philosophische Studien*, 1901, 17, pp 576–623.

Sensation and Perception in the History of Experimental Psychology, Edwin G. Boring, Appleton-Century-Crofts, New York, 1942.

'Human taste response as a function of locus of stimulation on the tongue and soft palate', Virginia B. Collings, *Perception and Psychophysics*, Vol. 16, January 1974, pp 169–174.

'The biological basis of food perception and acceptance', Linda M. Bartoshuk, *Food Quality and Preference*, Vol. 4, 1993, pp 21–32.

'Making sense of taste', David V. Smith and Robert F. Margolskee, *Scientific American*, March 2001, pp 32–39.

'Classical blunders', Linda Bartoshuck, Association for Psychological Science, *Observer*, March 2003, https://www.psychologicalscience.org/observer/classical-blunders

'The cells and logic for mammalian sour taste detection', Angela L. Huang et al., *Nature*, 24 August 2006, pp 934–938.

'The Tongue Map: Tasteless myth debunked', Christopher Wanjek, *Live Science*, 29 August 2006, https://www.livescience.com/7113-tongue-map-tasteless-myth-debunked.html

'Human taste thresholds are modulated by serotonin and noradrenaline', Tom P. Heath et al., *The Journal of Neuroscience*, 6 December 2006.

'Why do we like the taste of protein?', Jonah Lehrer, *Wired*, 23 June 2011, https://www.wired.com/2011/06/why-do-we-like-the-taste-of-protein/

'Science of umami taste: Adaptation to gastronomic culture', Kumiko Ninomiya, *Flavour*, 26 January 2015, http://www.flavourjournal.com/content/4/1/13

Balloon Popping

'Popping balloons: A case study of dynamical fragmentation', Sebastien Moulinet and Mokhtar Add-Bedia, *Physical Review Letters*, 30 October 2015, pp 184301-1–184301-5.

'Focus: Two modes of balloon bursting revealed', Philip Ball, *APS Physics*, 30 October 2015, https://physics.aps.org/articles/v8/105

Asteroid Dodging

'ESA plans mission to smallest asteroid ever visited', 4 February 2019, http://www.esa.int/Our_Activities/Space_Engineering_Technology/Hera/ESA_plans_mission_to_smallest_asteroid_ever_visited

'A plan to knock an asteroid off course', Deborah Byrd, 5 February 2019, *Earth and Sky*: https://earthsky.org/space/aida-didymoon-

plan-to-deflect-asteroid?utm_
source=EarthSky+News&utm_
campaign=f0969b5bd6-EMAIL_
CAMPAIGN_2018_02%E2%80%A6

Coming Clean on Clothes

'Georges Vigarello, Concepts of
cleanliness: Changing attitudes in France
since the Middle Ages', review by Roy
Porter, *Medical History*, April 1989, p 258.

'Georges Vigarello, Concepts of
cleanliness: Changing attitudes in
France since the Middle Ages', review
by Mathew Ramsey, *The American
Historical Review*, February 1991, pp
175–176.

'History of Washing Machines up to
1800', *Old and Interesting*, 14 April
2001, http://www.oldandinteresting.com/
history-washing-machines.aspx

'Size-dependent control of colloid
transport via solute gradients in dead-
end channels', by Sangwoo Shin,
Proceedings of the National Academy,
12 January 2016, pp 257–261.

'Focus: Rinsing is key to removing
stains', Michael Schirber, *APS Physics*,
16 March 2018, https://physics.aps.org/
articles/v11/28

'Cleaning by surfactant gradients:
Particulate removal from porous
materials and the significance of rinsing
in laundry detergency', Sangwoo Shin et
al., *Physical Review Applied*, 16 March
2019, pp 034012–1–034012–6.

'Diffusiophoresis found to be critical
factor for getting clothes clean', Bob
Yirka, Phys.Org, 23 March 2018, https://
phys.org/news/2018–03-diffusiophoresis-
critical-factor.html

'Coming clean: The physics of doing the
laundry', Phil Dooley, *Cosmos*, 23 March
2018, https://cosmosmagazine.com/
physics/coming-clean-the-physics-of-
doing-the-laundry

'Clothes washing mystery solved by
physicists', Michael Allen, *Physics World*,
3 April 2018, https://physicsworld.com/a/
clothes-washing-mystery-solved-by-
physicists/

Why Elephants Don't Get More Cancers

'Cancer and ageing in mice and men',
R. Peto et al., *British Journal of Cancer*,
October 1975, pp 311–426.

'Dark Angel: It defends you from disease
and decay, but its life-giving powers
come at a price. Can we turn the table
on this all-powerful gene?', David Lane,
New Scientist, 18 December 2004,
https://www.newscientist.com/article/
mg18424781–600-the-gene-with-your-
life-in-its-hands/

'Peto's paradox and the promise of
comparative oncology', Leonard Nunney
et al., *Philosophical Transactions of
the Royal Society B*, 19 July 2015, 370:
20140177.

'The multiple facets of Peto's paradox:
a life-history model for the evolution of
cancer suppression', by Joel S. Brown
et al., *Philosophical Transactions of
the Royal Society B*, 19 July 2015, 370:
20140221.

'Solutions to Peto's paradox revealed
by mathematical modelling and cross-
species cancer gene analysis', by Aleah
F. Caulin et al., *Philosophical Transactions
of the Royal Society B*, 19 July 2015, 370:
20140222.

'A metabolic perspective of Peto's
paradox and cancer', by Chi V. Dang,
*Philosophical Transactions of the Royal
Society B*, 19 July 2015, 370: 20140223.

'Comparative oncology: What dogs and
other species can teach us about humans
with cancer', Joshua D. Schiffman
and Matthew Brennan, *Philosophical
Transactions of the Royal Society B*,
19 July 2015, 370: 20140231.

'Towards cancer-aware life-history
modelling', Hanna Kokko and Michael E.
Hochberg, *Philosophical Transactions of
the Royal Society B*, 19 July 2015, 370:
20140234.

'Peto's paradox and human cancers',
Robert Noble et al., Philosophical
Transactions of the Royal Society B,
19 July 2015, 370: 20150104.

'Inclusive fitness effects can select
for cancer suppression into old age',
Joel S. Brown and C. Athena Aktipis,
*Philosophical Transactions of the Royal
Society B*, 19 July 2015, 370: 20150160.

'TP53 copy number expansion correlates
with the evolution of increased body size
and an enhanced DNA damage response
in elephants', by Michael Sulak et al.,
eLife, 16 October 2015, https://www.
biorxiv.org/content/10.1101/028522v3

'Elephants armed with genes that stop
cancer', *New Scientist*, 17 October
2015, p 19.

'Evolutionary adaptations to risk of
cancer: Evidence from cancer resistance
in elephants', Lisa M. Abegglen et
al., *Journal of the American Medical
Association*, 3 November 2015, pp
1806–1807.

'Potential Mechanisms for cancer
resistance in elephants and comparative
cellular response to DNA damage in
humans', Lisa M. Abegglen et al., *Journal
of the American Medical Association*,
3 November 2015, pp 1850–1860.

Cannibalism – A Nutritious Dining Option?

'Talk of cannibalism', Jared M. Diamond,
Nature, 7 September 2000, pp 25–26.

'Biochemical evidence of cannibalism at a
prehistoric Puebloan site in southwestern
Colorado', Richard A. Marlar et al.,
Nature, 7 September 2000, pp 74–78.

'Cannibalism and prion disease may
have been rampant in ancient humans',
Elizabeth Pennisi, *Science*, 11 April 2003,
pp 227–228.

'Balancing selection at the prion protein
gene consistent with prehistoric kurulike
epidemics', Simon Mead et al., *Science*,
11 April 2003, pp 640–643.

'The caring side of cannibalism', Andy
Coghlan, *New Scientist*, 28 November
2007, https://www.newscientist.com/
article/mg19626323–800-the-caring-
side-of-cannibalism/

'Corpse medicine: Mummies, cannibals and vampires', Richard Sugg, *The Lancet*, 21 June, 2008, pp 2078–2079.

'How Neanderthals met a grisly fate: Devoured by humans', Robin McKee, *The Guardian*, 17 May 2009, https://www.theguardian.com/science/2009/may/17/neanderthals-cannibalism-anthropological-sciences-journal

'Cultural cannabilism as a paleoeconomic system in the European Lower Pleistocene: The case of level TD6 of Gran Dolina (Sierra de Atapuerca, Burgos, Spain)', Eudald Carbonell et al., *Current Anthropology*, August 2010, pp 539–549.

'Human meat just another meal for early Europeans?', James Owen, National Geographic News, 2 September 2010, https://news.nationalgeographic.com/news/2010/08/100831-cannibalism-cannibal-cavemen-human-meat-science/

'What does a human taste like?', Ryan Mandelbaum, *Scientific American*, 19 December 2016, https://www.scientificamerican.com/article/what-does-a-human-taste-like/?WT.mc_id=SA_TW_HLTH_NEWS

'Cannabilism study finds people are not that nutritious', Erika Engelhaupt, National Geographic News, 6 April 2017, https://news.nationalgeographic.com/2017/04/human-cannibalism-nutrition-archaeology-science/

'Ancient cannibals didn't eat just for the calories, study suggests', Nicholas St Fleur, *The New York Times*, 6 April 2017, https://www.nytimes.com/2017/04/06/science/cannibalism-human-body-calories.html

'Prehistoric cannibalism not just driven by hunger, study reveals', Nicola Davis, 6 April 2017, *The Guardian*, https://www.theguardian.com/science/2017/apr/06/prehistoric-cannibalism-not-just-driven-by-hunger-study-reveals

'Assessing the calorific significance of episodes of human cannibalism in the Paleolithic', James Cole, *Scientific Reports*, 6 April 2017, 7:44707. DOI: 10.1038/srep44707.

Barcode Invention

'Scanning the globe: The humble bar code began as an object of suspicion and grew into a cultural icon. Today it's deep in the heart of the FORTUNE 500', Nicholas Varchaver, CNN, 31 May 2004, https://money.cnn.com/magazines/fortune/fortune_archive/2004/05/31/370719/index.htm

'The rise of the barcode', Finlo Rohrer, BBC News, 7 October 2009, http://news.bbc.co.uk/2/hi/uk_news/magazine/8295052.stm

'Barcode birthday: 60 years since patent', Zoe Kleinman, BBC News, http://www.bbc.co.uk/news/technology-19849141

'N. Joseph Woodland dies at 91: Co-inventor of bar code', 14 December 2012, *The New York Times*, https://www.nytimes.com/2012/12/13/business/n-joseph-woodland-inventor-of-the-bar-code-dies-at-91.html

'Remembering Joe Woodland, the man who invented the bar code', Cade Metz, *Wired*, 21 December 2012, https://www.wired.com/2012/12/joe-woodland-bar-code/

'The history of the bar code', Gavin Weightman, 23 September 2015, https://www.smithsonianmag.com/innovation/history-bar-code-180956704/

'Barcode Jesus flipbook', Scott Blake, https://www.youtube.com/watch?v=WDxCEdvn_A8

Alkaline Diet

'Acid/alkaline theory of disease is nonsense', Gabe Mirkin, MD, Quackwatch, 11 January 2009, https://www.quackwatch.org/01QuackeryRelatedTopics/DSH/coral2.html

'Meta-analysis of the effect of the Acid-Ash Hypothesis of osteoporosis on calcium balance', Tanis R. Fenton et al., *Journal of Bone and Mineral Research*, Vol. 24, No. 11, 2009, pp 1,835–1,840.

'When is a "milkshake" not a milkshake?' Ray Paulick, Horse Racing News, Paulick Report, 25 May 2012, https://www.paulickreport.com/news/ray-s-paddock/when-is-a-milkshake-not-a-milkshake/

'Systematic review of the association between dietary acid load, alkaline water and cancer', by Tanis R. Fenton and Tian Huang, *BMJ Open*, 2016; 6: e010438. DOI: 10.1136/bmjopen-2015-010438.

'Recent developments in the use of sodium bicarbonate as an ergogenic aid', L.R. McNaughton et al., *Current Sports Medical Reports*, July–August 2016, pp 233–244.

'The dying officer treated for cancer with baking soda', Dr Giles Yeo and Tristan Quinn, BBC News, 19 January 2017, https://www.bbc.com/news/magazine-38650739

'Alkaline diets: good for your health, or just another fad?', Tegan Taylor, *ABC Health and Wellbeing*, 1 August 2018, https://www.abc.net.au/news/health/2018-08-01/alkaline-diet-health-fact-or-fad/10055714?utm_source=sfmc&utm_medium=email&utm_campaign=%5brn_sfm%E2%80%A66

'Is alkaline water a miracle cure – or BS? The science is in', Arwa Mahdawi, *The Guardian*, 20 October 2018.

'More, please! A victim of cancer quack Robert O. Young wins a $105 million settlement', David Gorski, 5 November, 2018, https://www.theguardian.com/global/2018/oct/29/alkaline-water-cure-bs-science-beyonce-tom-brady?utm_

Fat Birds

'Empirical evidence for differential organ reductions during trans-oceanic bird flight', Phil F. Battley et al., *Proceedings of the Royal Society B*, 22 January 2000, pp 191–195.

'Wader birds off to Siberia', Janet Parker, ABC Science, 1 March 2003, http://www.abc.net.au/science/articles/2000/03/01/2685071.htm

'Body composition and flight ranges of bar-tailed godwits (Limosa lapponica baueri) from New Zealand', Phil F. Battley and Theunis Piersma, *The Auk*, 1 July 2005, pp 922–937.

'Extreme endurance flights by landbirds crossing the Pacific Ocean: Ecological corridor rather than barrier?', Robert E. Gill Jr et al., *Proceedings of the Royal Society B*, 21 October 2008, pp 447–457.

Pregnancy Double Whammy

'Superfetation in beef cattle', Joel Andrew Carter, a dissertation submitted to the Graduate Faculty of the Louisiana State University and Agricultural and Mechanical College in partial fulfillment of the requirements for the degree of Doctor of Philosophy in the Department of Animal and Dairy Sciences, December 2002.

'Superfoetation: à propos d'un cas et revue de la littérature (Superfetation: Case report and review of the literature)', O. Pape et al. *Journal of Gynecology Obstetrics and Human Reproduction* (formerly known as: *Journal De Gynécologie Obstétrique Et Biologie De La Reproduction*), December 2008, DOI: 10.1016/j.jgyn.2008.06.004.

'How can a pregnant woman get pregnant again', Dan Fletcher, *Time*, 28 September 2009, http://content.time.com/time/health/article/0,8599,1926414,00.html

'Superconception in mammalian pregnancy can be detected and increases reproductive output per breeding season', Kathleen Roellig et al. *Nature Communications*, 21 September 2010, DOI: 10.1038/ncomms1079.

'I got pregnant while I was already pregnant! Woman gives birth to two babies on the same day but they are NOT twins', Ashley Van Sipma, 1 July 2011, Mail OnLine, https://www.dailymail.co.uk/health/article-2010005/Woman-gives-birth-babies-day-NOT-twins.html

'Superfetation: Pregnant while already pregnant', Khalil A. Cassimally', *Scientific American Guest Blog*, 27 April 2011, https://blogs.scientificamerican.com/guest-blog/superfetation-pregnant-while-already-pregnant/

' "Hole in one, maybe", Woman falls pregnant TWICE in 10 days in rare medical phenomenon – and she only had sex once', Sarah Burns, *The Sun*, 14 November 2016, https://www.thesun.co.uk/living/2183219/woman-falls-pregnant-twice-in-10-days-in-rare-medical-phenomenon-and-she-only-had-sex-once/

'Pregnancy upon pregnancy', C. Claiborne Ray, *The New York Times*, 27 February 2017, https://www.nytimes.com/2017/02/27/science/superfetation-pregnancy.html?smid=tw-nytimesscience&smtyp=cur&_r=0

Spaghetti Snapping

'Dynamic buckling and fragmentation in brittle rods', J.R. Gladden et al., *Physical Review Letters*, 29 January 2005, pp 035503–1–035503–4, https://www.irphe.fr/~fragmix/publis/GHB2005a.pdf

'That's the way the spaghetti crumbles', Peter Weiss, *Science News*, 12 November 2005, p 315

'Vaulter's pole snaps into three pieces', BBC Sport, 8 August 2008, https://www.bbc.co.uk/sport/av/olympics/19179092

'Lazaro Borges (CB) snaps pole – pole vault: London 2012 Olympics. https://www.youtube.com/watch?v=VrHiK1aHWL0

'Here's how to bend spaghetti to your will', Maria Temming, *Science News*, 27 August 2018, https://www.sciencenews.org/article/heres-how-bend-spaghetti-your-will?utm_source=email&utm_medium=email&utm_campaign=latest-newsletter-v2

'Controlling fracture cascades through twisting and quenching', Ronald H. Heisser et al., *Proceedings of the National Academy of Sciences*, 28 August 2019, pp 8665–8670.

Vaping & E-Cigarettes – the Old Bait-and-Switch

'Vaping is Big Tobacco's bait and switch', Jeneen Interlandi, *The New York Times*, 8 March 2019, https://www.nytimes.com/2019/03/08/opinion/editorials/vaping-ecigarettes-nicotine-safe.html?emc=edit_th_190311&nl=todayshead-lines&nlid=183247910311

'Reducing the dangers of e-cigarettes for children: Opportunities for regulation and consumer education', Ryan D. Kennedy and Vanya C. Jones, *Medical Journal of Australia*, 18 February 2019, pp 118, 199.

'Regulating known unknowns: Ensuring the safety of e-liquids in Australia', Sarah L. White et al. *Medical Journal of Australia*, 18 February 201, pp 199, 120.

'Exposures to e-cigarettes and their refills: Calls to Australian Poisons Information Centres, 2009–2016', Carol Wylie et al. *Medical Journal of Australia*, 18 February 2019, p 126.

'Nicotine and other potentially harmful compounds in "nicotine-free" e-cigarette liquids in Australia', Emily Chivers et al. *Medical Journal of Australia*, 18 February 2019, pp 127, 128.

A2 vs A1 Milk

'A2 accuses dairy giant of suppressing milk defects', Deborah Hill Cone, 1 November 2002, http://www.sharechat.co.nz/article/4d8b820a/a2-accuses-dairy-giant-of-suppressing-milk-defects.html

'The A2 milk case: A critical review', A.S. Truswell, *European Journal of Clinical Nutrition*, 2 May 2005, pp 623–631.

'Review of the potential health impact of beta-casomorphins and related peptides', Report of the DATEX Working Group on beta-casomorphins, 29 January 2009, EFSA (European Food Safety Authority) Scientific Report (2009), 231, 1–107.

'Science or snake oil: Is A2 milk better for you than regular cow's milk?', Dr Nicholas Fuller, The Conversation, 23 August 2016, https://theconversation.com/science-or-snake-oil-is-a2-milk-better-for-you-than-regular-cows-milk-62486

'Digesting dairy: What's the difference between A2 and ordinary milk?', Tyne Logan, ABC WA Country Hour, 22 November 2017, http://www.abc.net.au/news/rural/2017-11-22/a2-milk-vs-

ordinary-milk/9177020?sf173960211=1
Australian government website for
health information for parents and
health professionals, https://www.
raisingchildren.net.au

'Lactation and breast cancer risk:
a case-control study in Connecticut',
T. Zheng, T.R. Holford, S.T. Mayne et al.
Br J Cancer 2001; 84:1472.

'Breastfeeding and reduced risk of breast
cancer in an Icelandic cohort study',
L.Tryggvadóttir, H. Tulinius, J.E. Eyfjord,
T. Sigurvinsson, *American Journal of
Epidemiology*, 2001; 154:37.

Kiss the Sun

'A journey to touch the Sun', Martin
Grolms, *Advanced Science News*,
29 August 2018, https://www.
advancedsciencenews.com/a-journey-to-
touch-the-sun/

'Parker Solar Probe changed the game
before it even launched', Rob Garner,
NASA, 5 October, 2018, https://www.
nasa.gov/feature/goddard/2018/parker-
solar-probe-changed-the-game-before-
it-even-launched

Do Fish Drink Water?

'Salt regulation in freshwater and
seawater fishes', Anja Soklic, 16
June 2017, https://blogionik.org/
blog/2017/06/16/salt-regulation/

'The multifunction fish gill: Dominant
site of gas exchange, osmoregulation,
acid-base regulation and excretion of
nitrogenous waste', David H. Evans et al.,
Physiology Review, 2005, Vol. 85,
pp 97–177.

'How is oxygen "sucked out" of
our waterways?', Stuart Khan, The
Conversation, 14 January 2019, http://
theconversation.com/how-is-oxygen-
sucked-out-of-our-waterways-109795

Human Faults

*Human Errors: A Panorama of Our
Glitches, from Pointless Bones to Broken
Genes*, Nathan H. Lents, Weidenfeld &
Nicolson, London, 2018.

Missing Mass of an Atom

'Standard model gets right answer for
proton, neutron masses', Ron Cowen,
Science News, 20 November 2008,
https://www.sciencenews.org/article/
standard-model-gets-right-answer-
proton-neutron-masses

'The weight of the world is quantum
chromodynamics', Andreas S. Kronfield,
Science, 21 November 2008, pp
1198–1199.

'Ab initio determination of light hadron
masses', S. Durr et al., *Science*, 21
November 2008, pp 1224–1227.

'Higgs found', Alexandra Witze, Science
News, 4 July 2012, https://www.
sciencenews.org/article/higgs-found

'Ab initio calculation of the neutron-
proton mass difference', Sz. Borsanyi
et al., *Science*, 27 March 2015, pp
1452–1455.

'The most precise picture of the proton',
DESY News, 2 July 2015, http://www.
desy.de/news/news_search/index_eng.
html?openDirectAnchor=829

'There's still a lot we don't know about
the proton', Emily Conover, *Science
News*, 18 April 2017, https://www.
sciencenews.org/article/theres-still-lot-
we-dont-know-about-proton

'Proton mass decomposition from the
QCD energy momentum tensor', Yi-Bo
Yang et al., *Physical Review Letters*, 19
November 2018, 212001–1 to 212001–6.

'Physicists finally calculated where
the proton's mass comes from', Emily
Conover, *Science News*, 26 November
2018, https://www.sciencenews.org/
article/proton-mass-quarks-calculation

Rogue Planets

'Bound and unbound planets abound',
Joachim Wambsganss, *Nature*, 19 May
2011, pp 289–290.

'Unbound or distant planetary mass
population detected by gravitational
microlensing', 'The Microlensing
Observations in Astrophysics
(MOA) Collaboration & The Optical

Gravitational Lensing Experiment (OGLE)
Collaboration', *Nature*, 19 May 2011, pp
349–352.

'New rogue planet found, closest to our
solar system', Nancy Atkinson, *Universe
Today*, 14 November 2012, https://www.
universetoday.com/98478/new-rogue-
planet-found-closest-to-our-solar-system/

'The hunt for rogue planets just got
tougher', Ramin Skibba, *Nature News*,
8 February 2017, https://www.nature.
com/news/the-hunt-for-rogue-
planets-just-got-tougher-1.21445?WT.
ec_id=NEWS-20170209&spMailingID=
53386858&spUserID=ODU3NzA%E2%
80%A6

'Fifty billion planets in our Milky Way
galaxy are likely to be free floaters, says
new study', Bruce Dorminey, *Forbes*,
4 March 2019, https://www.forbes.com/
sites/brucedorminey/2019/03/04/fifty-
billion-planets-in-our-milky-way-galaxy-
are-likely-to-be-free-floaters-says-new-
study/#8cf935851e75

'There may be 50 billion free-floating
planets in our galaxy', Paul Scott
Anderson, 10 March 2019, *EarthSky*,
https://earthsky.org/space/50-billion-
free-floating-planets-in-milky-way

'Survivability of planetary systems in
young and dense star clusters', A. van
Elteren et al. *Astronomy & Astrophysics*, 12 March 2019, https://arxiv.org/
pdf/1902.04652.bth_190311&nl=today-
sheadlines&nlid=183247910311

PICTURE CREDITS

All collages and illustrations by Pilar Costabal except where listed otherwise.
All images of Dr Karl Kruszelnicki by Steve Baccon.
NOTE: While efforts have been made to trace and acknowledge all copyright holders, in some cases these may have been unsuccessful. If you believe you hold copyright in an image, please contact the publisher.

References are to page numbers.

Why is Wombat Poo Cube-Shaped?
7   Wombat by FiledIMAGE/iStock.com
8   Fatso the Fat-Arsed Wombat by Saberywyn/Wikimedia Commons
9   Wombat road sign by Dutchnatasja/iStock.com
9   Wombat colon by Lisa Reidy

Closest Planet to Earth – And the Winner Is …
12   *Bottom:* 3D illustration of Mars by Jurik Peter/Shutterstock.com
14   Image of Sun, Mercury, Venus, Earth and Mars by Pixabay
15   Rainforest with the Sun behind by khlongwangchao/Shutterstock.com
16   *Top:* 3D illustration of Mars landscape by Dotted Yeti/Shutterstock.com
16   *Bottom:* A graphical representation of Mercury's internal structure by NASA's Goddard Space Flight Center

Genetic Testing Saves Babies
19   *Bottom:* Pipette and test tubes by ESB Professional/Shutterstock.com
21   Kids in a huddle by Steve Debenport/iStock.com

How Do Birds Sleep?
23   Robins by Max Forgues/Shutterstock.com
24   Cat with bird by Ornitolog82/iStock.com
25   Sleeping budgerigar by tsivbrav/iStock.com
25   Female rock ptarmigan by rockptarmigan/iStock.com
25   Sleeping ducks by undefined undefined/iStock.com
26   Brains (modified) by BodyParts3D, © 2008 Integrated Database Center for Life Science licensed by CC Show-Inheritance 2.1 Japan; commons.wikimedia.org/wiki/File:Corpus_callosum.png
27   *Top:* Black-tailed godwits in flight at Bhigwan by Lensalot/iStock.com
27   *Bottom:* Bar-tailed godwit by Denja1/iStock.com

Ships and Ocean-level Rise
29   Melting glacier in Antarctica by Bernhard Staehli/Shutterstock.com
31   Container ship at sea by Federico Rostagno/iStock.com

Hole Phobia?
33   Head of a lotus by Niwat.koh/Shutterstock.com
34   Blue-ringed octopus by Yusran Abdul Rahman/Shutterstock.com
34   Rash caused by coxsackie virus on child's hand by pavodam/Shutterstock.com
35   Nokia 9 camera lenses by Nokia

Big Brother Internet?
36   Collage by Pilar Costabal; man in wheelchair looking at phone by New Africa/Shutterstock.com; man with satchel by baona/iStock.com
37   Book cover of *1984* by George Orwell by Penguin Books
37, 38 and 42:
      Security camera 1 by Mark_Kuiken/iStock.com 531414905
43   'No vacancy' sign by Pgiam/iStock.com

Coral Spawning
44   Green turtle swimming on the Great Barrier Reef by Colin_Davis/iStock.com
44   Red and green zoanthids by chrisho/iStock.com
44   *Background:* Great Barrier Reef from helicopter by tororo/iStock.com
45   Parrot fish and bat fish by treetstreet/iStock.com
46   Sewer pipe flowing into the ocean by cmturkmen/iStock.com
47   Full moon over sea by Ig0rZh/iStock.com
48   Coral spawning underwater by Rich Carey/iStock.com
49   Tropical fish hiding in anemone by hypergurl/iStock.com

Tongue Taste Map
51   Woman eating from jar of spread by Adene Sanchez/iStock.com
51   Colourful sweets by MoMorad/iStock.com
52   Taste anatomy vector illustration diagram by normaals/iStock.com
52   Taste bud by Creative Commons, https://commons.wikimedia.org/wiki/File:1402_The_Tongue.jpg
52   Taste bud by Lisa Reidy
53   Champagne glasses by Poike/iStock.com
54   Portrait of young boy sticking out tongue by ozgurdonmaz/iStock.com
55   Figures from Dr David P. Hänig, 'Zur Psychophysik des Geschmackssinnes', *Philosophische Studien*, 17, 1901 via ECHO Cultural Heritage Online
56   'Distribution of tongue sensitivity along the edge of the tongue' graph in Edwin G. Boring, *Sensation and Perception in the History of Experimental Psychology*, 1942, Appleton-Century-Crofts, New York
57   Parmesan cheese on a wooden board by Olesia Shadrina/iStock.com

Balloon Popping
58   Boy blowing a bubble by stevecoleimages/iStock.com
59   Person doing bubble with chewing gum on bright background by Africa Studio/Shutterstock.com

Asteroid Dodging
61   Image of asteroid 2015 TB145 courtesy Creative Commons, commons.wikimedia.org/wiki/File:2015_TB145_discovery.gif
62   *Top and middle:* Diagrams, Creative Commons/Tomruen
62   *Bottom:* 2015 TB145 by NASA/JPL-Caltech/GSSR/NR40/AVI/NSF

Coming Clean on Clothes
65   Roman nobility women and slave, hand-coloured wood engraving, published c.1880 /iStock.com
66   Advertisement for Beetham's Royal Patent Washing Mill
68   Rack of second-hand jeans by Tendo23/iStock.com
69   Clothing in suds by deepblue4you/iStock.com
70   Indian fabric by blueclue/iStock.com
71   Child wearing odd socks by kissesfromholland/iStock.com

Why Elephants Don't Get More Cancers
72-77:
      Watercolour vector hand painted set with wild herbs and spices by Elena Medvedeva/dreamstime
74   An elephant's eye by sutlafk/iStock.com
75   Human cells by luismmolina/iStock.com

Cannibalism – A Nutritious Dining Option?
79   Vintage colour engraving of Hansel and Gretel by duncan1890/iStock.com
81   Parts of a doll's body in a sandwich by LoulouVonGlup/Shutterstock.com

PICTURE CREDITS

Barcode Invention

82, 82, 85, 88, 89 and 90:
  Barcode by Yamac Beyter/iStock.com

82   A man gets on the hip reader in operations directed on printed barcode. Warehouse scene by Anatoly Vartanov/Shutterstock.com

83   Vintage Pepsi-Cola by pixabay

83   Cut of beef by pixabay

83   Joseph Woodland by IBM

84   Lightbulb by actionvance/unsplash

84   Joseph Woodland's barcode patent courtesy US Patent and Trademark Office

85   Barcode Jesus Flipbook by Scott Blake by Scott Blake/Creative Commons

86   Customer grabbing a can off a supermarket shelf by Bill Selmeier/idhistory.com

86   Beanbag ashtray sleddogtwo/iStock.com

87   Circular barcode printing directions by Lisa Reidy

88   Warehouse worker by Tempura/iStock.com

88   Article on NCR 255 scanning system by NCR

89   Woman with shopping list by George Marks/iStock.com

Alkaline Diet

90   Digital bathroom scale by Seregraff/Shutterstock.com

91   Bicarbonate of soda by JPC-PROD/Shutterstock.com

95   Patient being rushed to operating room by Clerkenwell/iStock.com

96   Rainbow-coloured fruits and vegetables by Anna Shkuratova/Shutterstock.com

99   Test strip or water analysis on glass with water by vchal/Shutterstock.com

Fat Birds

101 Overweight young man by ozgurdonmaz/iStock.com

102 Bar-tailed godwit in flight by Dennis Jacobsen/Shutterstock.com

102 Flight paths of bar-tailed godwits by Robert E. Gill et al/USGS/Royal Society

103 Man in tight shirt by Lisa S./Shutterstock.com

104 Red knots feeding in the surf by Elliotte Rusty Harold/Shutterstock.com

Pregnancy Double Whammy

107 Swamp wallaby by Andrew Haysom/iStock.com

109 European brown hares in summer, Germany by Red Squirrel/Shutterstock.com

109 Julia Grovenburg's ultrasound

109 Kittens behind cat flap by Nils Jacobi/iStock.com

110 First triplets in the world to be conceived nearly two weeks apart by Barcroft Media/Getty Images

111 Human sperm and egg cells by koya979/Shutterstock.com

Fire in the Snow – Incinerating Toilet

113 Photos of incinerating toilet in Antarctica courtesy of the author

Spaghetti Snapping

114 Lazaro Borges at 2012 Olympic Games in London by Quinn Rooney/Getty Images

115 Modern glass building by ispyfriend/iStock.com

116 Broken glass texture by bjdlzx/iStock.com

116 Chef breaking spaghetti with his hands by ugurv/Shutterstock.com

117 Cooking spaghetti in a pot with boiling water by dibettadifino/Shutterstock.com

117 Sea waves by Can Inelli/iStock.com

118 Messy spaghetti baby by jfairone/iStock.com

118 Custom spaghetti-breaking device by Ronald Heisser and Vishal Patil courtesy Jorn Dunkel/MIT

119 Dragon sculpture on wall by FrameAngel/Shutterstock.com

Vaping and E-cigarettes – The Old Bait-and-Switch

121 Cigarette on hospital bed by Es sarawuth/Shutterstock.com

122, 125, 126:
  Purple photomicrograph of cancer cells by BonD80/Shutterstock.com

123 The face of vaping young man by master1305/iStock.com

124 X-ray of lungs with cancer by Hong xia/Shutterstock.com

125 Angel statue by Fedorenko/Shutterstock.com

126 Cancer cell among healthy cells by Shutterstock.com

127 Skull in cigarette tobacco by grafvision/Shutterstock.com

A2 vs A1 Milk

129 Herd of cows in a green field by symbiot/Shutterstock.com

132 Farmer milking a cow by Zacchio/Shutterstock.com

135 Cow-milking facility by yadamons/Shutterstock.com

Kiss the Sun

142 Rocket launching by NASA

143 Illustration of NASA's Parker Solar Probe approaching the Sun by NASA/Johns Hopkins APL/Steve Gribben

146 Parker Solar Probe prepping to launch for the Sun courtesy NASA/Johns Hopkins APL/Ed Whitman

146 Parker Solar Probe Mission by NASA, blogs.nasa.gov/parkersolarprobe/2018/08/11/the-parker-solar-probe-mission/

147 Parker Solar Probe 'tickets' courtesy of the author

Do Fish Drink Water?

151 Dead fish in contaminated lake by GroanGarbu/Shutterstock.com

154 and 155:
  Side view of a porkfish by GlobalP/iStock.com

155 Roach, river fish isolated on white background by Regfer/Shutterstock.com

Human Faults

158 Anatomy of an eye by solar22/iStock.com

158 Sinuses by kowalska-art/iStock.com

159 Voice nerves (modified), Creative Commons commons.wikimedia.org/wiki/File:Recurrent_laryngeal_nerve.svg

159 *Background:* Old Paper by Yellowdesignstudio/dreamstime

160 Hip joint by TefiM/iStock.com

160 Knee joint isolated on white by dlewis33/iStock.com

Missing Mass of an Atom

163 Schoolboy with arm raised in class by courtneyk/iStock.com

164 Atom with abstract scientific background by Pobytov/iStock.com

165 An atom made up of moving fire-like balls by zoom-zoom/iStock.com

167 Atom by zoom-zoom/iStock.com

169 Richard Phillips Feynman by 12/Universal Images Group via Getty Images

Rogue Planets

176 *Top:* View of the Orion Nebula via Hubble Space Telescope by NASA

176 *Bottom:* Closer view of Trapezium star cluster in Orion Nebula by NASA

177 Starry night by MD-TDH/Shutterstock.com

THANK YOU!

Navigating to the truth is the first step on any journey. So for help with stories in this brightly coloured volume I thank Professor Stewart Truswell ('A2 vs A1 Milk'), Professor Clare Collins ('A2 vs A1 Milk' and 'Alkaline Diet'), Dr Kyra Sim ('A2 vs A1 Milk' and 'Alkaline Diet'), Dr Nicholas Fuller ('A2 vs A1 Milk' and 'Alkaline Diet'), Brian Dunning (from Skeptoid.com – 'Alkaline Diet'), Leigh Stark ('Big Brother Internet?'), Dr Phil Dooley ('Coming Clean on Clothes'), Professor Terry Hughes ('Coral Spawning'), and Professor Geraint Lewis ('Missing Mass of an Atom' and 'Rogue Planets'). They made wonderful suggestions. 👍

I am very grateful to both Dr Tony Monger and Almost-Dr Petr Lebedev, who, with their wonderfully general Mental Maps, read all the stories and gave them a tune-up. Just like mechanics have a steel toolbox, physicists have a mental toolbox. 🦕

For me, the facts are enough. I know I find the story fascinating, but I have to find the pathway to make it interesting to you, the beloved passenger! Dr Mary Dobbie has guided me along that pathway for decades. She has detoured me away from blind alleys, made everything fun, and has linked tangles of loosely coherent concepts into a smoothly flowing superhighway of information – with fascination, awe and wonder as sign posts. 🎤😗

My ABC Radio producers then turn a script-for-reading into a script-for-listening-to – very different vehicles. Joanna Khan, Bernie Hobbs, Carl Smith, James Bullen, Dan Driscoll and Tiger Woods torque it up, and come in with a punchline every week. 🎤

Isabelle Benton (besides being my producer) has a special ability to match the script to the length of the trip – including the ten-year-olds-with-a-sense-of-curiosity. 🌷

Then comes the final lap, with the text being handed to the publisher. Thanks to Lu Sierra, who really brought the words home (editing is a very different skill from writing), and to Scott Forbes for additional editorial management. 🪣

This book looks very different from all my previous books – and I **love** it to pieces. This is thanks to designer Lisa Reidy and to Pilar Costabal, who crafted all the magnificent illustrations. 🖌️

Proofreading is the final pit stop. We all get together in one big room, and read every word on every page, looking for mistakes.

These grease monkeys included Chris Stedman (love a fact checker), Carmel Dobbie (love a school teacher), Isabelle Benton, Mary Dobbie, Jeanne Ryckmans and Jude McGee. 🤓

And, finally, the augmented reality. It's a bit of a gamble, but I'm quietly confident that Paul Kouppas (from Augmented Reality) can make it work. Almost-Dr Petr Lebedev shot the movies. 🎥

This fine book with ABC Books came about because of the good work of my agent, Jeanne Ryckmans. It's lovely coming back to what feels like home, with Jude McGee as my publisher. 📚

Of course, if nobody knows about the book, nobody will buy it. So Big Thanks to Matthew Howard (campaign manager), Larissa Bricis (digital content producer) and Steve Baccon (photographer). 💬

This book delivers truckloads of stories directly from my heart to you. 🚚🤍

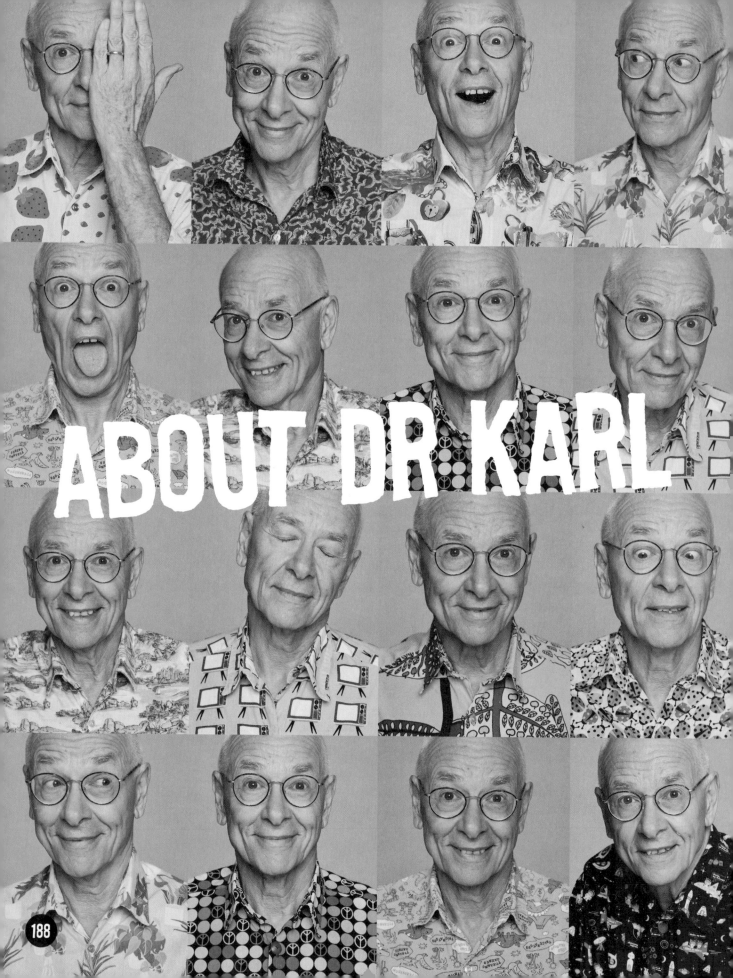

ABOUT DR KARL

Dr Karl Kruszelnicki AM loves science to pieces, and has been spreading the love in print, on TV and radio and online for more than thirty years.

The author of 45 books, Dr Karl is a lifetime student with degrees in physics and mathematics, biomedical engineering, medicine and surgery. He has worked as a physicist, labourer, roadie for bands, car mechanic, filmmaker, biomedical engineer, taxi driver, TV weatherman, and medical doctor at the Children's Hospital in Sydney.

Since 1995, he has been the Julius Sumner Miller Fellow at the University of Sydney. He has several weekly radio Q&A science shows in Australia including on triple j, and on the BBC in the UK.

He also does two free science Q&A sessions every Wednesday afternoon with schools around the world. (They're via the Internet. Check out drkarl.com to book one.)

GET EXTRA EASTER EGGS WITH THIS BOOK!

There are Easter eggs hidden inside this book! Let me explain.

I don't mean decorated eggs from Easter. Although, while we're here, did you know that decorating eggs goes back 60,000 years to Africa? Modern Easter eggs began some 2,000 years ago with the Christians of Mesopotamia, who stained eggs with red dye in memory of the blood of Christ. Today, chocolate Easter eggs are often hidden for children to find on Easter morning.

The other kind of Easter eggs were created in 1979. Warren Robinett was a programmer who developed the video game *Adventure*. His employer, Atari, didn't allow their programmers to be credited, fearing that competitors would poach their employees. Warren wanted acknowledgement, so he secretly inserted the message 'Created by Warren Robinett' into his game. If you found a specific pixel (the Gray Dot) and hovered your mouse over it during a certain part of the game, Warren's message would pop up.

Warren Robinett didn't tell Atari what he'd done, and later left the company. Then his secret message was accidentally found by a player. Due to cost, Atari decided not to remove the message. In fact, the company encouraged its programmers to insert more hidden messages for gamers to find. They called them 'Easter eggs'. (It turns out Warren might not have been the first Easter egger – a few earlier, similar Easter eggs have been found since – but Warren's is the most famous. AND it depends on what you mean by 'Easter egg'.)

For a while, Google Maps had an Easter egg. If you wanted directions from New York to Tokyo, Google Maps would tell you to kayak across the Pacific Ocean.

So, as I was saying, there are Easter eggs in this book. Of course, the book stands alone. But for hidden surprises, go to my homepage, drkarl.com, and follow the prompts. And if the Easter eggs don't pop up, at least you can still go Old School and read the book. It's not broken!

Happy Hunting!

For hidden surprises, go to my homepage, drkarl.com and follow the prompts.

The ABC 'Wave' device is a trademark of the
Australian Broadcasting Corporation and is used
under licence by HarperCollins*Publishers* Australia.

First published in Australia in 2019
by HarperCollins*Publishers* Australia Pty Limited
ABN 36 009 913 517
harpercollins.com.au

HarperCollins*Publishers*
Level 13, 201 Elizabeth Street, Sydney, NSW 2000, Australia
Unit D1, 63 Apollo Drive, Rosedale, Auckland 0632, New Zealand
A 53, Sector 57, Noida, UP, India
1 London Bridge Street, London, SE1 9GF, United Kingdom
Bay Adelaide Centre, East Tower, 22 Adelaide Street West, 41st Floor, Toronto, Ontario, M5H 4E3, Canada
195 Broadway, New York, NY 10007, USA

A catalogue record for this book is available
from the National Library of Australia

ISBN: 978 0 7333 4032 1 (paperback)
ISBN: 978 1 4607 1166 8 (ebook: epub)

Cover design by Lisa Reidy
Cover images: collage by Pilar Costabal; Dr Karl by Steve Baccon; car by Keith Bell/shutterstock.com
Endpapers by Pilar Costabal and Lisa Reidy
Internal design by Lisa Reidy
Colour reproduction by Splitting Image
Printed and bound in China by RR Donnelley

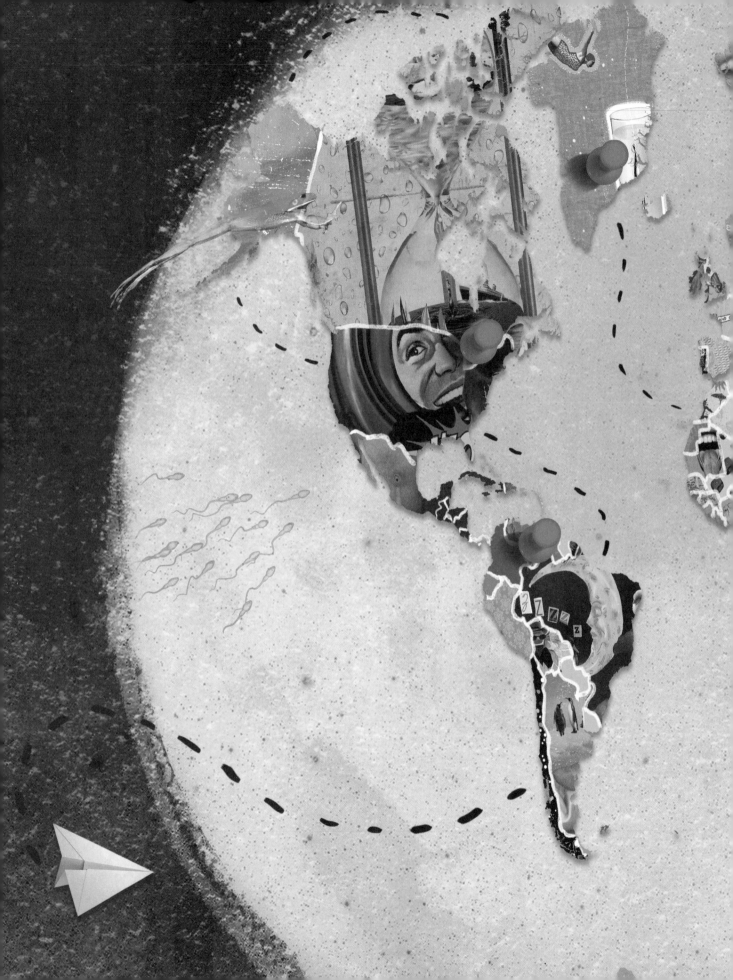